Nano-Electromagnetic Communication at Terahertz and Optical Frequencies

Related titles on electromagnetic waves:

Dielectric Resonators, 2nd Edition Kajfez and Guillon
Electronic Applications of the Smith Chart Smith
Fiber Optic Technology Jha
Filtering in the Time and Frequency Domains Blinchikoff and Zverev
HF Filter Design and Computer Simulation Rhea
HF Radio Systems and Circuits Sabin
Microwave Field-Effect Transistors: Theory, design and application, 3rd Edition Pengelly
Microwave Semiconductor Engineering White
Microwave Transmission Line Impedance Data Gunston
Optical Fibers and RF: A natural combination Romeiser
Oscillator Design and Computer Simulation Rhea
Radio-Electronic Transmission Fundamentals, 2nd Edition Griffith, Jr
RF and Microwave Modeling and Measurement Techniques for Field Effect Transistors
Jianjun Gao
RF Power Amplifiers Albulet
Small Signal Microwave Amplifier Design Grosch
Small Signal Microwave Amplifier Design: Solutions Grosch
2008+ Solved Problems in Electromagnetics Nasar
Antennas: Fundamentals, design, measurement, 3rd Edition Blake and Long
Designing Electronic Systems for EMC Duff
Electromagnetic Measurements in the Near Field, 2nd Edition Bienkowski and Trzaska
Fundamentals of Electromagnetics with MATLAB®, 2nd Edition Lonngren *et al.*
Fundamentals of Wave Phenomena, 2nd Edition Hirose and Lonngren
Integral Equation Methods for Electromagnetics Volakis and Sertel
Introduction to Adaptive Arrays, 2nd Edition Monzingo *et al.*
Microstrip and Printed Antenna Design, 2nd Edition Bancroft
Numerical Methods for Engineering: An introduction using MATLAB® and computational electromagnetics Warnick
Return of the Ether Deutsch
The Finite Difference Time Domain Method for Electromagnetics: With MATLAB® simulations Elsherbeni and Demir
Theory of Edge Diffraction in Electromagnetics Ufimtsev
Scattering of Wedges and Cones with Impedance Boundary Conditions Lyalinov and Zhu
Circuit Modeling for Electromagnetic Compatibility Darney
The Wiener–Hopf Method in Electromagnetics Daniele and Zich
Microwave and RF Design: A systems approach, 2nd Edition Steer
Spectrum and Network Measurements, 2nd Edition Witte
EMI Troubleshooting Cookbook for Product Designers Andre and Wyatt
Transmission Line Transformers Raymond Mack and Jerry Sevick
Electromagnetic Field Standards and Exposure Systems Grudzinski and Trzaska
Practical Communication Theory, 2nd Edition Adamy
Complex Space Source Theory of Spatially Localized Electromagnetic Waves Seshadri
Electromagnetic Compatibility Pocket Guide: Key EMC facts, equations and data Wyatt and Jost
Antenna Analysis and Design Using FEKO Electromagnetic Simulation Software Elsherbeni, Nayeri and Reddy
Scattering of Electromagnetic Waves by Obstacles Kristensson
Adjoint Sensitivity Analysis of High Frequency Structures with MATLAB® Bakr, Elsherbeni and Demir
Developments in Antenna Analysis and Synthesis vol. 1 and vol. 2 Mittra
Advances in Planar Filters Design Hong
Post-Processing Techniques in Antenna Measurement M. Castañer and L.J. Foged

Nano-Electromagnetic Communication at Terahertz and Optical Frequencies

Principles and Applications

Edited by
Akram Alomainy, Ke Yang, Muhammad A. Imran,
Xin-Wei Yao and Qammer H. Abbasi

The Institution of Engineering and Technology

Published by SciTech Publishing, an imprint of The Institution of Engineering and Technology, London, United Kingdom

The Institution of Engineering and Technology is registered as a Charity in England & Wales (no. 211014) and Scotland (no. SC038698).

The Institution of Engineering and Technology
Michael Faraday House
Six Hills Way, Stevenage
Herts, SG1 2AY, United Kingdom

www.theiet.org

British Library Cataloguing in Publication Data
A catalogue record for this product is available from the British Library

ISBN 978-1-78561-903-8 (hardback)
ISBN 978-1-78561-904-5 (PDF)

Typeset in India by MPS Limited
Printed in the UK by CPI Group (UK) Ltd, Croydon

Contents

Part II   Current development in THz components and interfaces   55

4   Terahertz antenna design for wearable applications   57
Abdel Baset, Muhammad Ali Imran, Akram Alomainy
and Qammer H. Abbasi

5   Terahertz (THz) application in food contamination detection   77
Aifeng Ren, Adnan Zahid, Xiaodong Yang,
Akram Alomainy, Muhammad Ali Imran and Qammer H. Abbasi

9 Error-control mechanisms for nano-electromagnetic communication networks   171

Xin-Wei Yao, De-Bao Ma and Chong Han

10 Conclusion and future outlook   195

Akram Alomainy, Ke Yang, Xin-Wei Yao, Muhammad Ali Imran and Qammer Hussain Abbasi

About the editors

Akram Alomainy (http://www.eecs.qmul.ac.uk/~akram/) received M.Eng. degree in communication engineering and Ph.D. degree in electrical and electronic engineering (specialized in antennas and radio propagation) from Queen Mary University of London (QMUL), UK, in July 2003 and July 2007, respectively. He joined the School of Electronic Engineering and Computer Science, QMUL, in 2007, where he is Associate Professor (Senior Lecturer) in the Antennas and Electromagnetics Research Group. He is a member of the Institute of Bioengineering and Centre for Intelligent Sensing at QMUL. His current research interests include small and compact antennas for wireless body area networks, radio propagation characterisation and modelling, antenna interactions with human body, computational electromagnetic, advanced antenna enhancement techniques for mobile and personal wireless communications, and advanced algorithms for smart and intelligent antenna and cognitive radio system.

Dr. Alomainy has secured various research projects funded by research councils, charities and industrial partners on projects ranging from fundamental electromagnetics to wearable technologies with a portfolio of around £4m as PI and Co-I. He is the lead of Wearable Creativity research at Queen Mary University of London and has been invited to participate at the Wearable Technology Show 2015, Innovate UK 2015 and also in the recent Wearable Challenge organised by Innovate UK IC Tomorrow as a leading challenge partner to support SMEs and industrial innovation. He has authored and co-authored 2 books, 6 book chapters and more than 250 technical papers in leading journals and peer-reviewed conferences.

Dr. Alomainy won the Isambard Brunel Kingdom Award, in 2011, for being an outstanding young science and engineering communicator. He was selected to deliver a TEDx talk about the science of electromagnetic and also participated in many public engagement initiatives and festivals. He was shortlisted twice in a row for 'Teacher of the Year' at Queen Mary University of London. He is a Chartered Engineer, member of the IET, senior member of IEEE, fellow of the Higher Education Academy (UK) and also a College Member for Engineering and Physical Sciences Research (EPSRC, UK) and its ICT prioritisation panels. He is also a reviewer for many funding agencies around the world including Expert Swiss National Science Foundation (SNSF) Research, the Engineering and Physical Sciences Research Council (EPSRC, UK) and the Medical Research Council (MRC, UK). He is an elected member of UK URSI (International Union of Radio

Science) panel to represent the UK interests of URSI Commission B (1 September 2014 until 31 August 2017).

Ke Yang received the Bachelor Degree in Electronic Information Engineering in 2009 and Master Degree in Electromagnetics Engineering in 2011 from Xidian University, China. He received his Ph.D. degree in Electronic and Electrical engineering from the Queen Mary University of London (QMUL), UK, in November 2015. Currently, he is Post-Doctoral Research Assistant at the Antenna and Electromagnetics group of QMUL. His research interests include nano-communication, Internet of things, millimeter and terahertz communication, body centric wireless communication issues and wireless body sensor networks.

Muhammad Imran (https://www.gla.ac.uk/schools/engineering/staff/muhammadimran/) is the Vice Dean Glasgow College UESTC and Professor of Communication Systems in the School of Engineering at the University of Glasgow. He was awarded his M.Sc. (Distinction) and Ph.D. degrees from Imperial College London, UK, in 2002 and 2007, respectively. He is an Affiliate Professor at the University of Oklahoma, USA, and a visiting Professor at 5G Innovation Centre, University of Surrey, UK. He has over 18 years of combined academic and industry experience, working primarily in the research areas of cellular communication systems. He has been awarded 15 patents, has authored/co-authored over 300 journal and conference publications, and has been principal/co-principal investigator of over £6 million sponsored research grants and contracts. He has supervised more than 30 successful Ph.D. graduates.

Professor Imran has an award of excellence in recognition of his academic achievements, conferred by the President of Pakistan. He was also awarded IEEE Comsoc's Fred Ellersick award 2014, FEPS Learning and Teaching award 2014 and Sentinel of Science Award 2016. He was twice nominated for Tony Jean's Inspirational Teaching award. He is a shortlisted finalist for The Wharton-QS Stars Awards 2014, QS Stars Reimagine Education Award 2016 for innovative teaching and VC's learning and teaching award in the University of Surrey. He is the co-editor of two books: *Access, Fronthaul and Backhaul Networks for 5G and Beyond* (IET, ISBN 9781785612138) and *Energy Management in Wireless Cellular and Ad-hoc Networks* (Springer, ISBN 9783319275666). He is a senior member of IEEE and a Senior Fellow of Higher Education Academy (SFHEA), UK.

Xin-Wei Yao is an Associate Professor at the College of Computer Science and Technology, Zhejiang University of Technology, China. He has been a visiting academic at the University at Buffalo, The State University of New York, USA, and Loughborough University, UK. His research interests are the design (protocols, architectures and applications) and analysis (modeling, simulation and implementation) of communication networks and bio-inspired networks. He has published 39 journals and conference papers in these areas, and has served on the organising and technical committees of several international conferences including the *International Conference on*

Quantum Nano/Bio, and Micro Technologies (ICQNM), International Conference on Cooperative Design, Visualization and Engineering (CDVE2), IEEE International Conference on Communications (ICC), International Conference on Recent Advances in Signal Processing, Telecommunications & Computing (SigTelCom), International Conference on Modeling, Simulation and Applied Optimization (ICMSAO) and *IEEE International Symposium on Personal, Indoor and Mobile Radio Communications (IMRC).* He is a reviewer of several international journals.

Qammer H. Abbasi (https://www.gla.ac.uk/schools/engineering/staff/qammer-abbasi/) received his B.Sc. and M.Sc. degrees in Electronics and Telecommunication Engineering from the University of Engineering and Technology (UET), Lahore, Pakistan (with distinction). He received his Ph.D. degree in Electronic and Electrical Engineering from Queen Mary University of London (QMUL), UK, in January 2012. From January 2012 to June 2012, he was Post-Doctoral Research Assistant at the Antenna and Electromagnetics group, QMUL, UK. From 2012 to 2013, he was an international young scientist under National Science Foundation China (NSFC) and Assistant Professor in the University of Engineering and Technology (UET), KSK, Lahore. From August 2013 to April 2017, he was with the Center for Remote Healthcare Technology and Wireless Research Group, Department of Electrical and Computer Engineering, Texas A&M University (TAMUQ), initially as Assistant Research Scientist and later was promoted as Associate Research Scientist and Visiting Lecturer, where he led multiple Qatar National Research Foundation grants (worth $3 million). Currently, Dr. Abbasi is Lecturer (Assistant Professor) at the University of Glasgow School of Engineering in addition to being Visiting Research Fellow with Queen Mary University of London (QMUL) and Visiting Associate Research Scientist with Texas A&M University (TAMUQ).

Dr. Abbasi has research portfolio of around $3 million and contributed to a patent, 5 books and more than 130 leading international technical journals and peer reviewed conference papers and received several recognitions for his research. Dr. Abbasi is an IEEE senior member and was Chair of IEEE young professional affinity group. He is an associate editor for *IEEE Journal of Electromagnetics, RF, and Microwaves in Medicine and Biology* and *IEEE Access* and acted as a guest editor for numerous special issues in top notch journals. He is a member of IET and committee member for *IET Antenna & Propagation* and healthcare network. Dr. Abbasi has been a member of the technical program committees of several IEEE flagship conferences and technical reviewer for several IEEE and top-notch journals. He contributed in organizing several IEEE conferences, workshop and special sessions in addition to European school of antenna course. His research interests include nano-communication, Internet of things, 5G and its applications to connected health, RF design and radio propagation, biomedical applications of millimetre and terahertz communication, wearable and flexible sensors, compact antenna design, antenna interaction with human body, implants, body centric wireless communication issues, wireless body sensor networks, non-invasive health care solutions, physical layer security for wearable/implant communication and multiple-input–multiple-output systems.

Chapter 1

Introduction to nano-communication

Akram Alomainy[1], Ke Yang[2], Xin-Wei Yao[3], Muhammad Ali Imran[4] and Qammer Hussain Abbasi[4]

In upcoming years, the advancement in nanotechnologies is expected to accelerate the development of integrated devices with the size ranging from one to a few hundred nanometers [1,2]. With the aim of developing miniaturised classical machines and creating nano-devices with new functionalities, nanotechnologies have produced and continued creating some novel nano-materials and nano-particles with new behaviours and properties that are not observed at the microscopic level. The links and connectivity between nano-devices distributed through collaborative efforts lead to the envision of nano-networks, followed by the nano-communication proposal. The limited capabilities of nano-machines in terms of processing power, complexity and range of operations can be expanded by this collaborative communication. It is sustaining the revolutionary transition from the Internet of things to the Internet of nano-things [3].

1.1 What is nano-communication?

According to Feynman, there is still plenty of room at the bottom [4]. Based on such statement and the considerable development of nanotechnology, it is pointed out by Professor Metin Sitti that the entire network systems would shrink into the nanoscale with the nano-robots and molecular machine as the elements in the near future [5]. With the introduction of nanotechnology, the idea of decreasing the size of the present sensor network into the nano level was also proposed, but at the same time, how to connect the nano-devices in such networks to conduct complex tasks was also questioned, leading to the proposal of the nano-network, followed by the introduction of the concept of nano-communication [6,7].

As the name indicates, nano-communication refers to the communication between nano-devices, which would adopt novel and modified communication and radio

[1]School of Electronic Engineering and Computer Science, Queen Mary University of London, London, UK
[2]School of Marine Science and Technology, Northwestern Polytechnical University, Xi'an, China
[3]College of Computer Science and Technology, Zhejiang University of Technology, Hangzhou, China
[4]James Watt School of Engineering, University of Glasgow, Glasgow, UK

propagation principles in comparison with conventional and existing solutions. To make it clearer, four requirements are summarised in IEEE P1906.1 [8] as follows:

- At least one essential component of the defined system should be at the nanoscale, even at just one dimension.
- The physical properties applied in the defined system should be different from the ones at the macro-scale. If we take the electromagnetic (EM) properties as an example, quantum effects can change it at the nanoscale. The resonant frequency of an antenna would no longer increase when its size decreases at the nanoscale; furthermore, the wave propagation velocity would be influenced leading to its reduction below the speed of light.
- The fundamentals of the communication theory should be mapped, where there should be a fully distinguishable transmitter, receiver, medium, message carriers and message.
- Some components of the proposed system should be artificial.

1.2   Envisioned communication methods of nano-communication

To connect the nano-devices, the communication between them needs to be completed. According to Akyildiz *et al.* [7], nano-communication can be divided into two scenarios: (1) communication between a nano-machine and a larger system such as the micro/macro-system and (2) communication between two or more nano-devices. Furthermore, the ways of EM, acoustic, nanomechanical, or molecular can all be applied to nano-communications [9], which will be discussed in this section.

1.2.1   Molecular paradigm

Molecular communication is considered as the most promising paradigm at the start of the nano era to achieve the nano-communication because there are numerous examples present in nature to learn and study. In molecular communication, an engineered miniature transmitter releases small particles into a propagation medium, while the molecules are applied to encode, transmit and receive information [10]. Molecular communication can be classified into several categories such as walkway-based (molecules propagate along a predefined pathway via molecular motors), flow-based (molecules propagate in a guided fluidic medium), diffusion-based (molecules propagate in a fluidic medium via spontaneous diffusion) and so on [2]. As the most common scheme in nature, the diffusion-based molecular communication (DMC) is most profoundly investigated in the literature: the mathematical framework for a physical end-to-end channel model for DMC [11], development of an energy model for DMC [12], modelling of diffusion noise [13], channel codes for reliability enhancement [14] and relaying-based solutions for increasing the range of DMC [15,16]. In contrast, the flow-based molecular communication is also studied, especially one of the communications in the circulatory system [17,18].

1.2.2 Electromagnetic paradigm

As the name indicates, EM methods use the EM wave as the carrier, and its properties such as amplitude, phase and delay are used to encode or decode the information. The possibility of EM communication is first investigated in [2] based on the findings that the terahertz (THz) band can be assigned as the operational frequency for future EM nano-transceivers, which can be made of the emerging new materials such as carbon nano-tube (CNT) and graphene [19]. In [20], the theoretical model of the nano-network whose nodes are made of the CNT was presented. Later, the channel model for THz wave propagating in the air with a different concentration of the water vapour was presented in [21], and the corresponding channel capacity was also studied. Based on the characteristics of the channel, a new physical layer aware medium access control protocol, time spread on-off keying (TS-OOK), was proposed in [22]. Meanwhile, the applications of THz technology in imaging and medical field [23,24] have also achieved great development, and the biological effects of THz radiation are reviewed in [25] showing a minimum effect on the human body and no strong evidence of hazardous side effects [10].

1.2.3 Acoustic paradigm

Acoustic propagation introduces slight pressure variations in the fluid or solid medium, which satisfy the wave equation. The behaviour of the nanorobots is relevant to their physical properties, surrounding medium and the working frequency. The feasibility of *in vivo* ultrasonic communication is evaluated by Hogg and Freitas [26], where communication effectiveness, power requirements and effects on the nearby tissue were examined on the basis of the discussion on the principles. Later, the nanoscale opto-ultrasonic communications in biological tissues were discussed in [27,28], where the generation and propagation model was studied, and in line with [26], the hazards and design challenges were investigated.

1.2.4 Other paradigms

Based on the development of the nanotechnology, a new paradigm of mechanical communication, that is touch communication (TouchCom), was also proposed in [17], where a swarm of nanorobots was used as message carriers for information exchange. In TouchCom, transient microbots (TMs) [29–31] were applied to carry the drug particles, which can be controlled and tracked by the external macro-unit (MAU) with a guiding force [18,32]. These TMs would survive some time in the body, and their pathway would be the channel for the information exchange while the process of loading and unloading is the corresponding transmitting and receiving process. The channel model of TouchCom was derived by defining the propagation delay, path loss with the angular/delay spectra of the signal strength [17]. Meanwhile, a simulation tool was proposed to characterise the movement of the nano-robot swarm in the blood vessel [32].

1.3   Development of nano-communication

Early studies in nano-communication were made in the 2000s by a limited group in the world, aiming to study the possibility of the nano-network, which put their efforts on the channel performance and the activity of the information carrier. The initial concept of nano-communication was first mentioned in 2004 by Cheng [33]. Later, the NaNoNetworking Center in Catalonia (N3Cat) has been set up in 2009 at the Universitat Politécnica de Catalunya, Spain, as an initiative of Professor Ian F. Akyildiz and Professor Josep Sol-Pareta with the main goals of carrying fundamental research studies on nano-networks, where three main projects were conducted: graphene-enabled wireless networks-on-chip for massive multicore architectures, graphene-enabled wireless communications and fundamentals and applications of molecular nano-networks through cell signalling [34]. Then, Nano Communications Centre was established at the Tampere University of Technology, Finland, to mainly study the molecular communication for the nano-network where bacterial nano-networks, neuronal networks and calcium signalling were investigated [35].

Recently, numerous groups spring up from 2010, aiming not only on the channel but also on the modulation, coding, routing and networking. In 2014, Dr Josep Miquel Jornet, whose PhD study topic was mainly on EM nano-communication at the THz band [36], built a group in the University at Buffalo, The State University of New York, aiming at the THz channel characterisation on the basis of the model and design of the graphene-based plasmonic nano-antennas and nano-transceivers. Professor Ian F. Akyildiz started a project on molecular nano-communication (MoNaCo) networks supported by the National Science Foundation. To expand the current work to body-centric communication, Dr Akram Alomainy started research studies on the theoretical and numerical studies with Dr Ke Yang on the channel performance of nano-communication at the THz band at the Queen Mary University of London [37–39]. Also, Dr Yansha Deng at King's College London and Dr Weisi Guo in the University of Warwick start the group to study the performance of the molecular communication [40,41].

1.4   Book organisation

The book is divided into three different parts: Part I (Chapters 2 and 3) deals with the fundamentals and state-of-the-art advances in nano-EM communication network, whereas Part II (Chapters 4 and 5) deals with the current development of THz technology, followed by Part III which summarises the advances in the physics and network layers of nano-EM communication; and in the end, some concluding remarks and the future of the nano-EM network are presented in Chapter 10.

Chapter 2 starts with the comparison of the metal antenna with the graphenna, followed by the description of the novel transceivers made by graphene, which is the elementary part of the EM communication. Then, the channel performance in the open air is discussed. Before the illustration of the network performance, the mechanism in the aspect of modulation, synchronisation and coding/decoding is discussed. In the end, the envisioned applications in the biomedical, environmental, security, defence and consumer fields are described in detail.

Chapter 3 envisions a targeted drug delivery system to describe the cross-scale structure of the nano-communication network. The system includes an external MAU and a number of *in vivo* drug-loaded micro-units, which combines both the EM way and molecular communication. A simulation platform is built to investigate the system performance under the framework of IEEE 1906.1 in the aspects of field, medium, message carriers, message, motion, perturbation and specificity. Furthermore, a corresponding experimental platform of the magnetic-driven bacteria targeted drug delivery (TDD) system is built to verify the model and conclusion from the simulation.

Chapter 4 deals with the antenna design technologies for THz body-centric application. After a detailed review on the design of graphene antenna and human effects on the traditional antenna radiation characteristics, a novel material, Pervoskite, is fully investigated due to its promising advantages such as superconductivity, ferroelectricity and low cost. In the end, a Perovskite THz antenna is designed and demonstrates its high potential for short-range THz communication.

Chapter 5 presents the possibility of THz sensing in food contamination detection based on the review of various sensing technologies. Additionally, current THz detection systems are illustrated, which would help the reader build the whole picture of THz detection systems. By analysing the important parameters, such as absorption coefficient and transmission response of different state of various fruits, THz sensing can be considered a promising candidate to change the current plant-monitoring techniques.

Chapter 6 discusses the channel performances with the inclusion of human tissues for the *in vivo* wireless nano-sensor networks. Following the traditional analysis method, the end-to-end transmission is first investigated in the aspects of the path loss model, noise model, information rate based on TS-OOK modulation, etc., followed by the discussion of the network performance of the multiuser scenario based on the interference model. The results provide an important theoretical basis more practically for network-level modelling and stimulate further research on simple, reliable and energy-efficient communication protocols and coding schemes.

Chapter 7 discusses the current state of modulation, coding and synchronisation issues faced by the nanoscale communication systems operating in the THz band. Based on a broad review of the techniques utilised in the microwave band and below, the possibility of the available schemes is investigated. In the modulation scheme, the pulse-based modulation methods are compared with the traditional carrier-based modulation, while for the coding scheme, the low-weight coding scheme is regarded as the promising one with the consideration of the peculiar characteristics of the THz communication. In the end, the synchronisation issues are discussed in detail from the aspects of device level and network level.

Chapter 8 focuses on the analysis of the existing routing protocols for wireless nano-networks (WNNs) to find appropriate routing protocols to guarantee multihop communication. The current routing protocols, such as RADAR routing, CORONA, SLR, LSDD, DEROUS, MHTD, EEMR, ECR and TEForward, can be classified into three categories: limit flood area-based routing protocols, dynamic infrastructure-based routing protocols and single path-based routing protocols. By a detailed comparison with each other, the general design code is introduced at the end of the chapter, where two guided rules of constrained resources and limited energy supply should always be reminded.

Chapter 9 discusses the error control mechanisms for nano-EM networks based on the performance evaluation of different error control schemes: automatic repeat request (ARQ), forward error correction, error prevention codes (EPCs) and a hybrid EPC. Based on the complete analysis, a novel error control mechanism is proposed by considering the trade-off between energy harvesting and consumption for perpetual nano-networks. Then, the energy state model based on the extended Markov chain approach of the proposed error control strategy with a probing mechanism is also validated and investigated by simulation and numeral calculations where the end-to-end successful packet delivery probability, end-to-end packet delay, achievable throughput and energy consumption are fully investigated and evaluated.

Chapter 10 discusses the road ahead for the nano-EM communication network in the aspect of the challenges and research trends.

References

[1]   Akyildiz IF, Brunetti F, and Blázquez C. Nanonetworks: A new communication paradigm. Computer Networks. 2008;52(12):2260–2279.

[2]   Akyildiz IF, and Jornet JM. Electromagnetic wireless nanosensor networks. Nano Communication Networks. 2010;1(1):3–19.

[3]   Akyildiz IF, Jornet JM, and Pierobon M. Nanonetworks. Communications of the ACM. 2011;54(11):84.

[4]   Feynman RP. There's plenty of room at the bottom. Engineering and Science. 1960;23(5):22–36.

[5]   Sitti M, Ceylan H, Hu W, *et al*. Biomedical applications of untethered mobile milli/microrobots. Proceedings of the IEEE. 2015;103(2):205–224.

[6]   Bush SF. Nanoscale Communication Networks. London: Artech House; 2010.

[7]   Akyildiz IF, Brunetti F, and Blázquez C. Nanonetworks: A new communication paradigm. Computer Networks. 2008;52(12):2260–2279.

[8]   Bush SF, Eckford A, Paluh J, *et al*. IEEE draft recommended practice for nanoscale and molecular communication framework. IEEE P19061/D11, October 2014. 2014. pp. 1–52.

[9]   Andrew AM. Nanomedicine, Volume 1: Basic capabilities. Kybernetes. 2000;29(9/10):1333–1340.

[10] Yang K. Characterisation of the in-vivo THz communication channel within the human body tissues for future nano-communicaiton networks. School of Electronic Engineering and Computer Science, Queen Mary University of London; 2016.

[11] Pierobon M, and Akyildiz IF. A physical end-to-end model for molecular communication in nanonetworks. IEEE Journal on Selected Areas in Communications. 2010;28(4):602–611.

[12] kr Kuran M, Yilmaz HB, Tugcu T, *et al.* Energy model for communication via diffusion in nanonetworks. Journal of Nano Communication Networks. 2010;1(2):86–95.

[13] Pierobon M, and Akyildiz IF. Diffusion-based noise analysis for molecular communication in nanonetworks. IEEE Transactions on Signal Processing. 2011;59(6):2532–2547.

[14] Shih PJ, Lee CH, Yeh PC, *et al.* Channel codes for reliability enhancement in molecular communication. IEEE Journal on Selected Areas in Communications. 2013;31(12):857–867.

[15] Einolghozati A, Sardari M, and Fekri F. Relaying in diffusion-based molecular communication. In: IEEE International Symposium on Information Theory (ISIT); 2013. pp. 1844–1848.

[16] Nakano T, and Liu JQ. Design and analysis of molecular relay channels: An information theoretic approach. IEEE Transactions on NanoBioscience. 2010;9(3):213–221.

[17] Chen Y, Kosmas P, Anwar P, *et al.* A touch-communication framework for drug delivery based on a transient microbot system. IEEE Transactions on Nanobioscience. 2015;14(4):397–408.

[18] Khalil ISM, Magdanz V, Sanchez S, *et al.* Magnetic control of potential microrobotic drug delivery systems: Nanoparticles, magnetotactic bacteria and self-propelled microjets. In: Engineering in Medicine and Biology Society (EMBC), 2013 35th Annual International Conference of the IEEE; 2013. pp. 5299–5302.

[19] da Costa MR, Kibis O, and Portnoi M. Carbon nanotubes as a basis for terahertz emitters and detectors. Microelectronics Journal. 2009;40(4): 776–778.

[20] Koksal CE, and Ekici E. A nanoradio architecture for interacting nano-networking tasks. Nano Communication Networks. 2010;1(1):63–75.

[21] Jornet JM, and Akyildiz IF. Channel modeling and capacity analysis for electromagnetic wireless nanonetworks in the terahertz band. IEEE Transactions on Wireless Communications. 2011;10(10):3211–3221.

[22] Jornet JM, Pujol JC, and Pareta JS. Phlame: A physical layer aware MAC protocol for electromagnetic nanonetworks in the terahertz band. Nano Communication Networks. 2012;3(1):74–81.

[23] Joseph CS, Yaroslavsky AN, Neel VA, *et al.* Continuous wave terahertz transmission imaging of nonmelanoma skin cancers. Lasers in Surgery and Medicine. 2011;43(6):457–462.

[24] Jung E, Park H, Moon K, *et al.* THz time-domain spectroscopic imaging of human articular cartilage. Journal of Infrared, Millimeter, and Terahertz Waves. 2012;33(6):593–598.

[25] Wilmink GJ, and Grundt JE. Invited review article: current state of research on biological effects of terahertz radiation. Journal of Infrared, Millimeter, and Terahertz Waves. 2011;32(10):1074–1122.

[26] Hogg T, and Freitas Jr RA. Acoustic communication for medical nanorobots. Nano Communication Networks. 2012;3(2):83–102.

[27] Santagati GE, and Melodia T. Opto-ultrasonic communications for wireless intra-body nanonetworks. Nano Communication Networks. 2014;5(1):3–14.

[28]   Santagati GE, and Melodia T. Opto-ultrasonic communications in wireless body area nanonetworks. In: 2013 Asilomar Conference on Signals, Systems and Computers. IEEE; 2013. pp. 1066–1070.

[29]   Hwang SW, Tao H, Kim DH, *et al.* A physically transient form of silicon electronics. Science. 2012;337(6102):1640–1644.

[30]   Martel S, Mohammadi M, Felfoul O, *et al.* Flagellated magnetotactic bacteria as controlled MRI-trackable propulsion and steering systems for medical nanorobots operating in the human microvasculature. International Journal of Robotics Research. 2009;28(4):571–582.

[31]   Martel S, Felfoul O, Mathieu JB, *et al.* MRI-based medical nanorobotic platform for the control of magnetic nanoparticles and flagellated bacteria for target interventions in human capillaries. International Journal of Robotics Research. 2009;28(9):1169–1182.

[32]   Chen Y, Kosmas P, and Wang R. Conceptual design and simulations of a nano-communication model for drug delivery based on a transient microbot system. In: 2014 8th European Conference on Antennas and Propagation (EuCAP). IEEE; 2014. pp. 63–67.

[33]   Kaifu C. Nano-photoelectronics devices in nano-communication. Nanoscience & Technology. 2004;3:008.

[34]   Homepage of NaNoNetworking Center in Catalonia at Universitat Politècnica de Catalunya, Spain. Available from: http://www.n3cat.upc.edu/index.

[35]   Homepage of Nano Communications Centre at Tampere University of Technology, Finland. Available from: http://et4nbic.cs.tut.fi/nanocom/index.html.

[36]   Jornet JM, and Akyildiz IF. Fundamentals of electromagnetic nanonetworks in the terahertz band. Foundations and Trends in Networking. 2013; 7(2–3): 77–233.

[37]   Yang K, Pellegrini A, Munoz MO, *et al.* Numerical analysis and characterization of THz propagation channel for body-centric nano-communications. IEEE Transactions on Terahertz Science and Technology. 2015;5(3).

[38]   Yang K, Pellegrini A, Brizzi A, *et al.* Numerical analysis of the communication channel path loss at the THz band inside the fat tissue. In: 2013 IEEE MTT-S International Microwave Workshop Series on RF and Wireless Technologies for Biomedical and Healthcare Applications (IMWS-BIO). IEEE; 2013. pp. 1–3.

[39]   Zhang R, Yang K, Abbasi QH, *et al.* Impact of cell density and collagen concentration on the electromagnetic properties of dermal equivalents in the terahertz band. IEEE Transactions on Terahertz Science and Technology. 2018;8(4):381–389.

[40]   Guo W, Deng Y, Li B, *et al.* Eavesdropper localization in random walk channels. IEEE Communications Letters. 2016;20(9):1776–1779.

[41]   Deng Y, Noel A, Elkashlan M, *et al.* Modeling and simulation of molecular communication systems with a reversible adsorption receiver. IEEE Transactions on Molecular Biology and Multi-Scale Communications. 2015;1(4):347–362.

Part I

Fundamentals and state-of-the-art advances in nano-electromagnetic communication network

Chapter 2

Fundamentals and applications of nano-electromagnetic communications

Chong Han[1]

Nanotechnology is providing the engineering community with a new set of tools to create miniature machines, which has a few cubic micrometers in size, and functions including sensing, actuation, computation, and data storing [1]. A large number of such nanomachines can accomplish more complex tasks collaboratively, with the capability of wireless communications. In light of this direction, nanonetworks, i.e., networks of nanomachines, can enable transformative and promising applications in the biomedical, environmental, security, defense, and consumer fields, as revealed by Akyildiz *et al.* [2].

2.1 Fundamentals of nano-electromagnetic communications

2.1.1 Nano-electromagnetic communications in the terahertz band

Enabling nanoscale electromagnetic (EM) communication relies on the development of nanoscale transceivers and antennas. Very high resonant frequencies would be imposed by the miniaturization of a metallic antenna. For example, a one-micrometer-long dipole antenna is expected to resonate at 150 THz, i.e., in the near-infrared spectrum. However, at infrared and visible optical frequencies, metals do not behave as perfect electric conductors anymore because the phase velocity of currents at higher frequencies for the case of metals is much below the speed of light. Therefore, the standard radio frequency antenna theory cannot be applied in this case, which results in the challenge that common assumptions in antenna theory need to be revised as mentioned by Dorfmuller *et al.* [3] and Nafari and Jornet [4], and it was proposed that antennas cannot be described by surface currents only at optical frequencies. As a result, we need to assume a constant volume current for the derivation of an analytical model. Moreover, in addition to the metallic

[1]University of Michigan–Shanghai Jiao Tong University Joint Institute, Shanghai Jiao Tong University, Shanghai, China

antennas, optical nano-transmitters (e.g., nano-lasers) and optical nano-receivers (e.g., nano-photodetectors) will be needed for nanoscale optical communications. At optical frequencies, the feasibility of nanonetworks would be compromised due to the very high propagation loss and the limited power of nanomachines.

To overcome the aforementioned limitations of metallic antenna and optical transceivers, an alternative technology of graphene can be utilized to develop novel nano-transceivers and nano-antennas, which is capable of resonating at lower frequencies than their metallic counterparts. Promisingly, graphene is a two-dimensional carbon crystal with extraordinary electrical conductivity to propagate extremely high-frequency electrical signals [1]. Specifically, graphene supports the propagation of surface plasmon polariton (SPP) waves, which are surface-confined EM waves that appear at the interface of a metal (graphene in this case) and a dielectric (air or any other substrate on which graphene is deposited), at room temperature [5,6].

Jornet and Akyildiz [7,8] first proposed the idea of utilizing graphene to build plasmonic nano-antennas for wireless communications. By leveraging the reduced propagation speed of SPP waves on graphene, the fundamental principle is to design resonant antennas, with a one micrometer long and tens to hundreds of nanometer wide, which can efficiently radiate at frequencies in the 0.1–10 THz band, i.e., at a frequency two orders of magnitude lower than metallic antennas with the same size. By following this direction, Tamagnone *et al.* [9] and Dragoman *et al.* [10] studied that the resonant frequency of the antennas can be electrically tuned and even considered different shapes and designs.

In a complete communication system, mechanisms to generate, modulate, detect, and demodulate the THz signals are needed, besides antennas. In Jornet and Akyildiz [11], a THz plasmonic signal source and a detector based on a hybrid graphene III–V semiconductor high electron mobility transistor (HEMT) were proposed. Compared with existing plasmonic sources and detectors, the generated plasma wave is not directly radiated but utilized to launch an SPP wave on the graphene layer that extends toward the nano-antenna. Other options include the use of photoconductive antennas, which downconvert optical radiation into THz-band radiation [12], plasmonic grating structures [13], or quantum cascade lasers [14]. All of these studies point out that the THz band is envisioned as the frequency band of nano-electromagnetic communications.

2.1.2   *Terahertz-band channel modeling*

Free-space propagation: Characterizing the THz channel is a necessary step toward enabling nanoscale THz communications. While the frequency regions immediately below (i.e., microwave and millimeter-wave) and above (i.e., infrared and visible optical) this band have been extensively investigated, this is one of the least-explored frequency bands in the EM spectrum. In Jornet and Akyildiz [15], the first channel model for the entire THz band was developed, by utilizing tools from radiative transfer theory, electromagnetics and communication theory, and leveraging the contents of the high-resolution transmission molecular absorption (HITRAN) database.

Two main phenomena affecting the propagation of EM signals at THz frequencies in free space are spreading and molecular absorption. The spreading loss, which is common to any wireless communication system, accounts for the attenuation due to the expansion of the wave as it propagates through the medium, and it depends only on the signal frequency and the transmission distance. The absorption loss accounts for the attenuation that a propagating wave suffers due to molecular absorption, i.e., the process by which a part of the wave energy is converted into internal kinetic energy in some of the molecules that are found in the medium. This depends on the concentration and the particular mixture of molecules encountered along the path. Different types of molecules have different resonant frequencies, and, in addition, the absorption at each resonance is not confined to a single frequency but spread over a range of frequencies. Consequently, the terahertz channel is very frequency selective. A list of absorption-defined spectral windows is provided in Figure 2.1, with carrier frequency f_c, 3-dB bandwidth B_{3dB}, molecular absorption loss A_{abs}, and total path loss A, over the distances at 0.1, 1, and 10 m. While for long distances, the effort is currently on characterizing individual transmission windows (e.g., 300 GHz [16]), it is clear that for nanoscale THz communications, the entire THz band can be considered as a single transmission window.

Multipath effects: The presence of obstacles further complicates the propagation of THz waves. Any object whose size is beyond a few wavelengths (i.e., a few millimeters or more) behaves as an obstacle for THz signals, which can reflect, scatter, or diffract THz waves leading to multipath propagation. In this direction, a unified multiray channel model for the THz band was first provided in Han *et al.* [17], based on ray-tracing techniques and accounting the propagation models for the line of sight (LoS), reflected, scattered, and diffracted paths. The developed

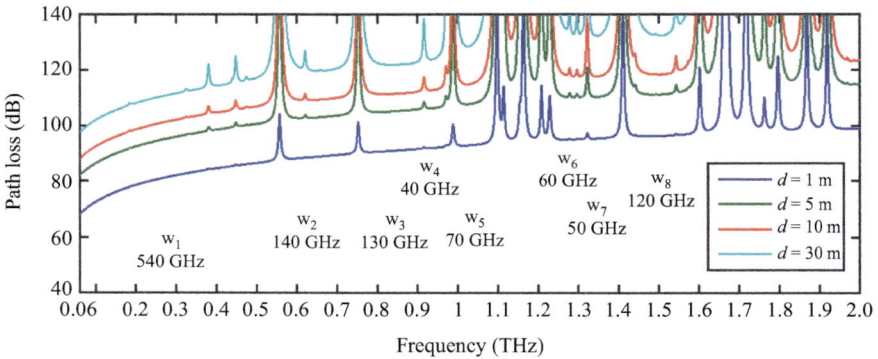

Figure 2.1   Path loss in the THz band. Eight spectral windows are identified, and their bandwidths range from 40 GHz up to 0.54 THz

theoretical model was validated with the experimental measurements at 60 and 300 GHz from the literature. Then, using the developed propagation models, an in-depth analysis on the THz channel characteristics is carried out, in terms of the distance-varying and frequency-selective spectral windows, coherence bandwidth, wideband channel capacity, and the temporal broadening effects. The principles to develop an efficient multiray model are summarized here. When LoS is available, the direct ray dominates the received signal energy, while the reflected rays play a dominant role when LoS is absent. As the operating frequency increases, the surface is seen to be rougher, and hence, more power is scattered out of the specular direction. Hence, scattered rays are very important and have to be included in the ray-tracing model for both LoS and NLoS conditions. Furthermore, the diffraction path can be ignored in general, only except when the receiver is in the very closed region near the incident shadow boundary. Further, stochastic modeling studies are presented by Hossain *et al.* [18], He *et al.* [19], and Han *et al.* [20].

2.1.3   Communication mechanism

Due to the hardware peculiarities of the nano-things and the application scenarios in which they will be used, there are several research challenges in the realization of the nanoscale communications that require innovative solutions and even to rethink some well-established concepts in communication and network theory, which includes understanding the communication channel behavior and accordingly the channel modeling, new modulation techniques, nanoscale synchronization, light-weight coding solutions, and beamforming techniques, and tailor new solutions to this new domain in order to increase their communication distances.

Modulation: The very large bandwidth supported by the THz-band channel over short distances enables new modulation strategies, which are also well suited for the limited capabilities and energy constraints of nanomachines. For example, ultra-low-energy sub-picosecond-long pulses can be utilized. These pulses, as shown in Figure 2.2, reserving major frequency components are between 0.5 and 4 THz, have already been widely used in THz spectroscopy and imaging systems. In this direction, a new communication scheme based on the transmission of one-hundred-femtosecond-long Gaussian pulses by following an on-off keying modulation spread in time was first proposed in Jornet and Akyildiz [21]. In this scheme, a logical "1" is transmitted using a one-hundred-femtosecond-long pulse and a logical "0" is transmitted as silence, i.e., the device remains silent when a logical zero is transmitted. To benefit from this transient behavior of molecular absorption and account for the temporal broadening effect, the time between symbols needs to be much longer than the pulse duration (e.g., a few tens of picoseconds), and, for convenience, it is a fixed parameter in the proposed scheme. Note that under this scheme, many nano-devices can simultaneously transmit without creating interference. Indeed, as the time between transmissions is expectedly much longer than the pulse duration, several nanomachines can concurrently use the channel without affecting each other. To fully exploit the phenomenon, low-weight channel coding strategies have been proposed [22].

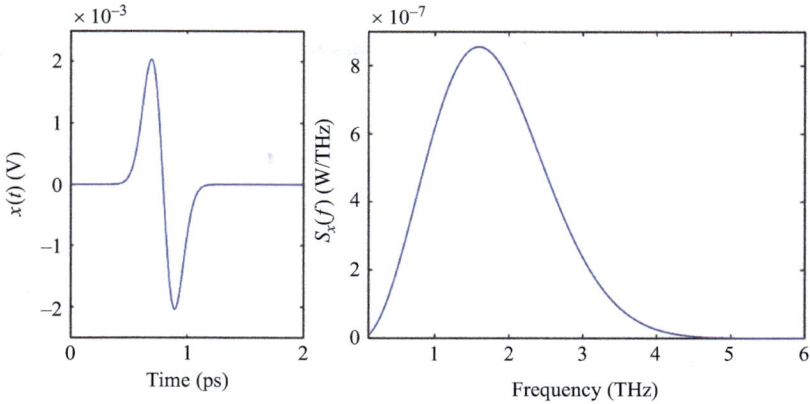

Figure 2.2 Time and frequency responses of a femtosecond-long pulse for THz-band communications

Furthermore, as an extension from the carrierless to the carrier-based modulation scheme, a multi-wideband waveform design for distance-adaptive THz-band communications is developed by Han *et al.* [23] and Akyildiz *et al.* [24], which includes the features of the pseudo-random time-hopping sequence and the polarity randomization. To cope with the unique characteristics and improve the distance, the pulse waveform design in the distance-adaptive multi-wideband system for the THz band includes the development of the waveform model and the dynamical adaptation of the rate and the transmit power on each sub-window. An optimization framework is formulated to solve these design parameters, with the aim of maximizing the communication distance while satisfying the rate and the transmit power constraints.

Synchronization: The huge bandwidth of the THz band comes at costs. First, a high-frequency- and distance-selective path loss causes severe distortion, including attenuation and temporal broadening effects on the transmitted pulses. Second, the digital synchronization, which has the advantages of cost efficiency, full integration, and robustness, requires multi-hundred-Giga-samples per second (Gs/s) and even Tera-samples per second (Ts/s) sampling rates, while the fastest sampling rate to date does not exceed 100 Gs/s. Due to these reasons, timing errors as small as picoseconds can seriously degrade the system performance. Therefore, accurate synchronization for these ultra-short pulse communications is critical, for which a low sampling rate (LSR, i.e., a fraction of Nyquist rate) timing acquisition algorithm is proposed, by extending the theory of sampling signals with a finite rate of innovation from compressive sampling in signal processing to the communication context. An LSR algorithm for timing acquisition is presented by Han *et al.* [25], which exploits the properties of the annihilating filter [26] and considers the THz communication parameters, which include the antenna gain, the distance, the number of frames per symbol, and the pulse width. The proposed LSR algorithm has high performance with uniform sampling at 1/20 of the Nyquist rate when the signal-to-noise ratio is high.

Coding and decoding: The channel peculiarities and the expected capabilities of transceivers require the development of novel channel codes for THz-band communication. The classical capacity approaching channel codes are designed to maximize the data rate for a given transmit power or equivalently to minimize the transmit power for a target data rate. However, in addition to transmit power, the decoding power is another fundamental source of power consumption. Indeed, for distances smaller than 10 m, the decoding power required by most state-of-the-art decoders is often comparable to, or even larger than, the transmit power. Uncoded transmission is commonly used in 60 GHz systems to reduce the decoding power, despite increasing the transmission power. Hence, classical error control mechanisms need to be revised before being used for nanoscale communication. On the one hand, automatic repeat request (ARQ) mechanisms might not be suited for nano-networks due to the energy limitations of nano-devices, which have very limited energy and require nanoscale energy-harvesting mechanisms to operate. On the other hand, the majority of forward error correction (FEC) mechanisms are too complex for the expected capabilities of the nano-devices. For this, Jornet [22] proposed for the first time the use of low-weight channel codes for error prevention in nanonetworks. These are able to simultaneously reduce molecular absorption noise and interference and, thus, prevent channel errors from happening beforehand.

More recently, Yao *et al.* [27] propose a novel error control strategy with probing (ECP) mechanism for the nanonetworks powered by energy harvesting. In particular, each data packet is transmitted only after the successful communication of one probing packet. Moreover, a probabilistic analysis of the overall network traffic and multiuser interference is used by the proposed energy state model to capture the dynamic network behavior. The performance of the ECP mechanism and the other four different error control strategies, namely, ARQ, FEC, error prevention codes (EPC), and a hybrid EPC, in terms of the end-to-end successful packet delivery probability, end-to-end packet delay, achievable throughput, and energy consumption, are investigated and evaluated. The observations include that the proposed ECP mechanism can maximize the end-to-end successful data packet delivery probability than the other four error control schemes, increase the achievable throughput compared with ARQ and EPC schemes, and outperform the ARQ and FEC schemes in terms of energy utilization.

2.1.4   Network protocols

Network architecture and medium access: The internetworking among the nano-things, the micro-devices, and the Internet remains an open issue, as illustrated in Figure 2.3. Furthermore, the design of routers and gateways for the Internet and the traffic bottleneck problem need to be revisited, by considering the massive amount of data in the nanonetwork realm. Two main architectures for nanonetworks have been considered in the recent literature, namely ad hoc nanonetworks and

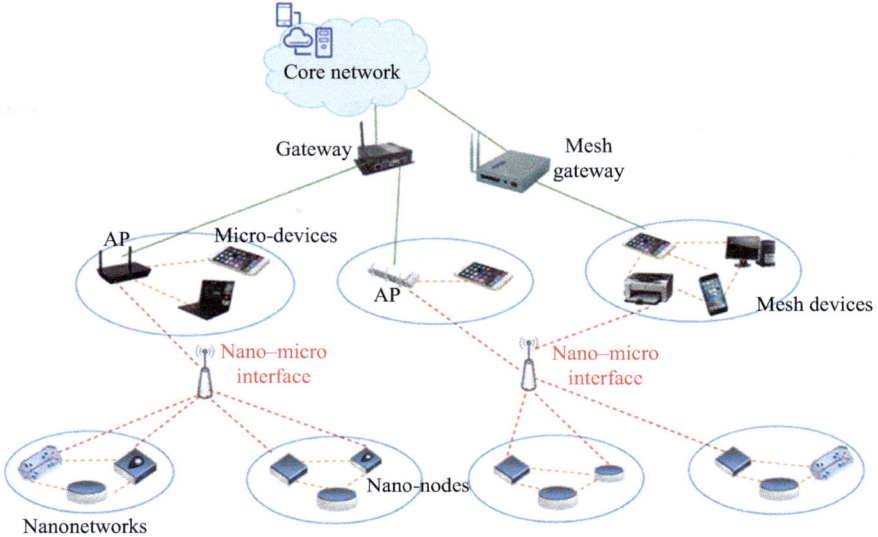

Figure 2.3   *Nanonetworks realize internetworking among the nano-things, the micro-devices, and the Internet*

infrastructure nanonetworks. In ad hoc nanonetworks, nanomachines interact directly with each other. The limitation of the existing medium access schemes arises. First, the use of a handshake effectively limits the throughput of nanonetworks when compared to the supported data rates in the THz band. Second, nanomachines might not have enough computational resources to dynamically find the optimal communication parameters as well as to handle such ultra-high data rates.

The characteristics of THz communications pose the following challenges, as described by Han *et al.* [28]. First, the narrow beams and directivity cause deafness problem, which prevent the alignment of transceivers. Second, the required data transmission coverage and deafness avoidance demands complicate the control channel selection mechanisms in node discovery and coupling process, which need to decide the adoption of either lower frequency or THz and the type of directionality, namely, omni, semi, or fully directional for the sending reception antenna mode. Third, LoS obstacles make the cell boundary/access point (AP) coverage defined by the received signal strength amorphous. Moreover, LoS blockage together with the user movements can disrupt link connectivity and pose challenges to link robustness. Finally, although the reduced multiuser interference (MUI) in THz communication facilitates the spatial reuse, the significant MUI, especially in a small range and dense network, poses requirements for interference monitoring and effective concurrent transmission scheduling.

By realizing these challenges, efforts have been made to design the efficient medium access control (MAC) protocol for THz communication networks. Among others, Han *et al.* [29] propose a memory-assisted MAC protocol with angular division multiplexing for centralized THz networks. In the network association phase, AP tests all the angular slots using continuous directional narrow beams, while the receiving nodes equipped with omni-directional antennas select the best angular slot. The memory of angular slots assists the AP in avoiding unnecessary scanning. Moreover, Xia *et al.* [30] propose a receiver-initiated synchronous MAC protocol for the THz network. The surrounding transmitters are assumed to point their narrow beams to the receiver. While the potential receiver sequentially scans the whole space to broadcast clear to send (CTS) frames, overcoming the deafness problem. The surrounding transmitter that has transmission demand will send DATA directionally after receiving the CTS. The receiver will send a positive acknowledgment frame once the transmission is successful. Otherwise, after a time-out, the transmitter will execute a random backoff process. Furthermore, by leveraging out of THz-band assistance, Tong and Han [31] propose a multi-radio-assisted MAC (MRA-MAC) scheme in distributed THz networks, where the neighbor discovery is executed by the angle of arrival estimation, and the network geometry information can be removed from the frame structure. A dual-band operation combining the 2.4/5 GHz and THz is performed in MRA-MAC. The shared same best path is assumed between the 2.4/5 GHz control communications and the THz data transmissions.

Interference: Interference and coverage analysis based on the stochastic geometry method are studied by Yao *et al.* [32], taking into account the nanonetwork density, THz frequency band, and interference strength. High density of APs with a small beam width and the transmission at THz frequencies with a low absorption efficient, such as 0.67 THz, are recommended to mitigate the interference and achieve a better coverage performance. Similarly, using the tools of stochastic geometry, Petrov *et al.* [33] study the systems operating in the THz band by explicitly capturing high directivity of the transmit and receive antennas, molecular absorption, and blocking of high-frequency radiation. Two radiation pattern models of directional antennas are considered, namely the cone model representing an ideal directional antenna and the cone plus sphere model capturing specifics of a nonideal directional antenna with side lobes. The metrics of interest are the mean interference and the signal-to-interference plus noise (SINR) ratio at the receiver. Observations are drawn for the same total emitted energy by a Poisson field of interferers, both the interference and the SINR significantly increase when simultaneously both transmit and receive antennas are directive. Furthermore, blocking has a profound impact on the interference and SINR creating much more favorable conditions for communications compared with no blocking case.

Addressing and routing: To connect a very large number of nano-things to the Internet, the addressing becomes an important issue. Although IPv6 addressing has been proposed for low-power wireless communication nodes, more efficient addressing schemes are expected to take the advantages of the hierarchical network architecture in the nanonetworks. Furthermore, Pierobon *et al.* [34] incorporate multi-hop transmissions from nanomachines to the nano-controller.

Compared with other existing hybrid architectures, it is the nano-controller which establishes the best approach for a nanomachine to follow, in light of the probability of energy savings, which depends on the channel properties, physical layer, and energy of the nanomachine. However, the impact of the computational resources on the proposed solutions is not captured. Independent of the architecture, device-aware and resource-efficient routing protocols solutions for nano-networks do not exist to date.

End-to-end reliability: Currently, no efforts have been made to guarantee for the end-to-end links, either from a remote command center to the nano-nodes or from the nano-things to a common sink. A cross-layer module is expected to lay out the fundamental part of an optimization framework to obtain the optimal routing paths and the communication parameters, by exploiting the inter-relations among different layer functionalities in the nanonetwork by Han *et al.* [35]. In particular, the cross-layer solutions need to accurately capture both the high heterogeneity and the unique features of the network architecture in the nanonetwork paradigm.

Security: Security and privacy are important issues in nanoscale communication. On the one hand, the nanoscale communication is vulnerable to attacks since the limited capability of the nano-devices cannot support sophisticated security schemes. Moreover, wireless links and unattended status increase the probability of compromising authentication and data integrity. On the other hand, the ways in which data collection, mining, and provisioning in the nanonetworks are completely different from the existing networks. The problem of privacy arises since the collection of personal information is very hard to be controlled. Traditionally, it is expected that eavesdropping becomes essentially impossible when the transmitted signal has sufficiently high directionality. The eavesdropper's techniques are to place a cylindrical object in the path of the transmission to scatter radiation toward the eavesdropper since cylindrical surfaces can scatter radiation over a wide range of angles. As a countermeasure, the transmitter can use a transceiver to detect and distinguish the differences between the backscattered signals from the cylindrical and the THz wireless channel.

2.2 Applications of nano-electromagnetic communications

- *In vivo* **health monitoring and drug delivery systems** (see Figure 2.4): Nanosensors can be implanted to operate inside the human body in real-time and provide fast and accurate disease diagnosis. Furthermore, autonomous nanorobots can be used to serve drug delivery and even minimally invasive surgery [36]. Using a nano-to-micro interface, these nano-devices can interact with larger micro-devices (e.g., a wearable device or the user's smartphone) to ultimately converge relevant information to a healthcare provider. Furthermore, THz radiation is nonionizing and therefore poses no known health risks to cells except for heating.

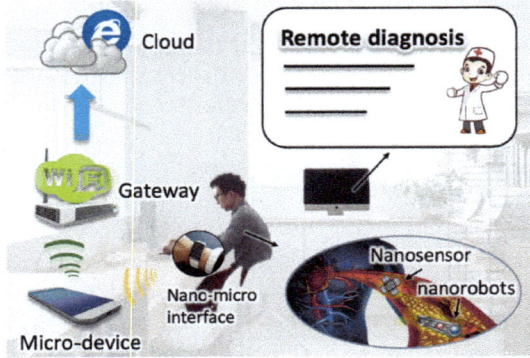

Figure 2.4 *Nanonetworks enabled* in vivo *health monitoring and drug delivery systems*

Figure 2.5 *Nanosensors for structure health monitoring*

- **Nanoscale structure health monitoring** (see Figure 2.5): Physical nano-sensors, operating in the THz frequency range, could be utilized to measure the physical integrity and detect induced damage in various dielectric, composite, and porous materials in the automotive and aerospace industries. These damages were estimated by sensing the difference between electromagnetic properties of these materials when exposed to heat damage as it was found that dielectric properties change when materials are exposed to heat or high temperatures. Dielectric properties are described by complex electrical permittivity ($\varepsilon = \varepsilon - i\varepsilon'$) and complex magnetic permeability ($\mu = \mu - i\mu'$), where the real part participates in shift in resonance frequency and the imaginary part participates in resistive losses [37]. The electromagnetic radiation using terahertz spectroscopy and terahertz pulsed imaging is a promising choice to achieve both high spatial resolution and acceptable penetration depth.

To cloud-based database

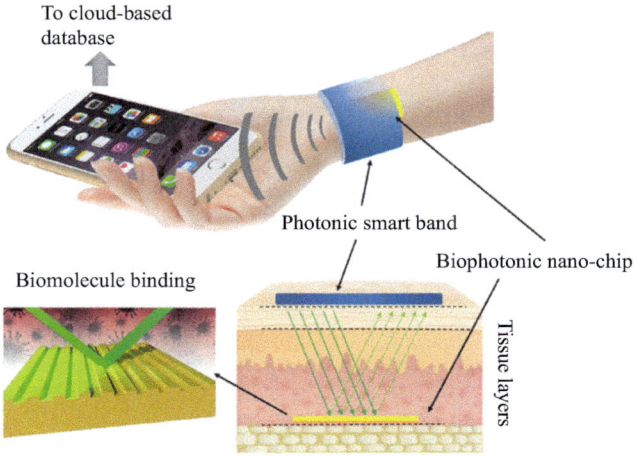

Photonic smart band

Biophotonic nano-chip

Biomolecule binding

Tissue layers

Figure 2.6   Wearable biosensing nanonetworks

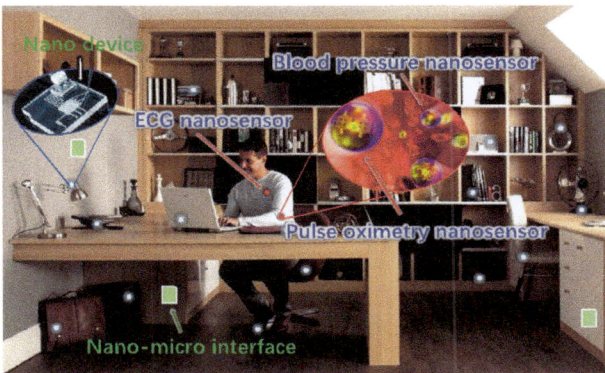

Nano device

Blood pressure nanosensor

ECG nanosensor

Pulse oximetry nanosensor

Nano-micro interface

Figure 2.7   The Internet of nano-things

- **Nuclear, biological, and chemical defenses** (see Figure 2.6): Chemical and biological nanosensors can be strategically placed in the walls or interleaved with the fabrics of security agents' clothing to detect harmful chemicals and biological weapons in a distributed manner and enhance the performance of chemical reactors. These nanosensors are not only several orders of magnitude smaller than existing sensors but also faster and much more accurate in terms of event detection, by precisely sensing the concentration as low as one molecule.
- **The Internet of nano-things** (see Figure 2.7): The interconnection of nano-scale machines with existing communication networks and ultimately Internet

Figure 2.8   Wireless network-on-chip (WiNoC) communications in the THz band

defines a truly cyber-physical system, which is known as the Internet of nano-things [38] as one step forward from the Internet of things. In an inter-connected office, a nano-transceiver and nano-antenna can be embedded in every single object to allow them to be permanently connected to the Internet. Consequently, a user can keep track of all professional and personal items in an effortless fashion.

- **Wireless network-on-chip communications** (see Figure 2.8): The THz band can provide efficient and scalable means of intercore communication in wire-less network-on-chip (WiNoC) networks, using planar nano-antennas to create ultra-high-speed links, which are more per the growing circuit integration in the semiconductor manufacturing. More importantly, the use of graphene-based THz-band communication [39] would deliver inherent multicast and broadcast communication capabilities at the core level. The significant dif-ference between the carrier frequency of the THz signals and the operating frequency of the digital circuits (under 5 GHz) alleviates the switching noise coupled from the digital circuits. Moreover, the THz signals suffering from high molecular absorption attenuation at resonant frequencies do not exist in the closed WiNoC environment [40].

References

[1]   Ferrari AC, Bonaccorso F, Fal'Ko V, *et al.* (2015) Science and technology roadmap for graphene, related two-dimensional crystals, and hybrid systems. Nanoscale 7(11):4598–4810.

[2]   Akyildiz IF, Jornet JM, and Pierobon M (2011) Nanonetworks: A new frontier in communications. Communications of the ACM 54(11):84–89.

[3]   Dorfmuller J, Vogelgesang R, Khunsin W, Rockstuhl C, Etrich C, and Kern K (2010) Plasmonic nanowire antennas: Experiment, simulation, and theory. Nano Letters 10(9):3596–3603.

[4] Nafari M, and Jornet JM (2017) Modeling and performance analysis of metallic plasmonic nano-antennas for wireless optical communication in nanonetworks. IEEE Access 5:6389–6398.

[5] Koppens FHL, Chang DE, and Garcia de Abajo FJ (2011) Graphene plasmonics: a platform for strong light matter interactions. Nano Letters 11(8): 3370–3377.

[6] Vakil A, and Engheta N (2011) Transformation optics using graphene. Science 332(6035):1291–1294.

[7] Jornet JM, and Akyildiz IF (2010) Graphene-based nano-antennas for electromagnetic nanocommunications in the terahertz band. In: Proc. of 4th European Conference on Antennas and Propagation, EUCAP, pp. 1–5.

[8] Jornet JM, and Akyildiz IF (2013) Graphene-based plasmonic nano-antenna for terahertz band communication in nanonetworks. IEEE Journal on Selected Areas in Communications 31(12):685–694.

[9] Tamagnone M, Gomez-Diaz JS, Mosig JR, and Perruisseau-Carrier J (2012) Reconfigurable terahertz plasmonic antenna concept using a graphene stack. Applied Physics Letters 101(21):214102.

[10] Dragoman M, Neculoiu D, Bunea AC, *et al.* (2015) A tunable microwave slot antenna based on graphene. Applied Physics Letters 106(15):153,101.

[11] Jornet JM, and Akyildiz IF (2014) Graphene-based plasmonic nano-transceiver for terahertz band communication. In: Proc. of European Conference on Antennas and Propagation (EuCAP).

[12] Yardimci NT, Yang SH, Berry CW, and Jarrahi M (2015) High-power terahertz generation using large-area plasmonic photoconductive emitters. IEEE Transactions on Terahertz Science and Technology 5(2):223–229.

[13] Otsuji T, Watanabe T, Boubanga Tombet S, *et al.* (2013) Emission and detection of terahertz radiation using two-dimensional electrons in III-V semi-conductors and graphene. IEEE Transactions on Terahertz Science and Technology 3(1):63–71.

[14] Lu Q, Wu D, Sengupta S, Slivken S, and Razeghi M (2016) Room temperature continuous wave, monolithic tunable THz sources based on highly efficient mid-infrared quantum cascade lasers. Scientific Reports 6:23595.

[15] Jornet JM, and Akyildiz IF (2011) Channel modeling and capacity analysis for electromagnetic wireless nanonetworks in the terahertz band. IEEE Transactions on Wireless Communications 10(10):3211–3221.

[16] Priebe S, Jastrow C, Jacob M, Kleine-Ostmann T, Schrader T, and Kurner T (2011) Channel and propagation measurements at 300 GHz. IEEE Transactions on Antennas and Propagation 59(5):1688–1698.

[17] Han C, Bicen AO, and Akyildiz IF (2015) Multi-ray channel modeling and wideband characterization for wireless communications in the terahertz band. IEEE Transactions on Wireless Communications 14(5):2402–2412.

[18] Hossain Z, Mollica C, and Jornet JM (2017) Stochastic multipath channel modeling and power delay profile analysis for terahertz-band communication. In: Proceedings of the 4th ACM International Conference on Nanoscale Computing and Communication, ACM, p. 32.

[19]   He D, Guan K, Fricke A, *et al.* (2017) Stochastic channel modeling for kiosk applications in the terahertz band. IEEE Transactions on Terahertz Science and Technology 7(5):502–513.

[20]   Han C, and Chen Y (2018) Propagation modeling for wireless communications in the terahertz band. IEEE Communications Magazine 56(6):96–101.

[21]   Jornet JM, and Akyildiz IF (2014) Femtosecond-long pulse-based modulation for terahertz band communication in nanonetworks. IEEE Transactions on Communications 62(5):1742–1754.

[22]   Jornet JM (2014) Low-weight error-prevention codes for electromagnetic nanonetworks in the terahertz band. Nano Communication Networks 5(1–2): 35–44.

[23]   Han C, Bicen AO, and Akyildiz IF (2016) Multi-wideband waveform design for distance-adaptive wireless communications in the terahertz band. IEEE Transactions on Signal Processing 64(4):910–922.

[24]   Akyildiz IF, Han C, and Nie S. (2018) Combating the distance problem in the millimeter wave and terahertz frequency bands. IEEE Communications Magazine 56(6):102–108.

[25]   Han C, Akyildiz IF, and Gerstacker WH (2017) Timing acquisition and error analysis for pulse-based terahertz band wireless systems. IEEE Transactions on Vehicular Technology 66(11):10102–10113.

[26]   Maravic I, and Vetterli M (2005) Sampling and reconstruction of signals with finite rate of innovation in the presence of noise. IEEE Transactions on Signal Processing 53(8):2788–2805.

[27]   Yao XW, Ma DB, and Han C (2019) ECP: A probing-based error control strategy for THz-based nanonetworks with energy harvesting. IEEE Access.

[28]   Han C, Zhang X, and Wang X (2019) On medium access control schemes for wireless networks in the millimeter-wave and Terahertz bands. Nano Communication Networks 19:67–80.

[29]   Han C, Tong W, and Yao X (2017) MA-ADM: A memory-assisted angular-division-multiplexing MAC protocol in Terahertz communication networks. Nano Communication Networks 13:51–59.

[30]   Xia Q, Hossain Z, Medley M, and Jornet J (2015) A link-layer synchronization and medium access control protocol for Terahertz-band communication networks. In: Proc. of IEEE GLOBECOM, pp. 1–7.

[31]   Tong W, and Han C (2017) MRA-MAC: a multi-radio assisted medium access control in Terahertz communication networks. In: IEEE GLOBECOM, pp. 1–6.

[32]   Yao XW, Wang CC, Wang WL, and Han C (2017) Stochastic geometry analysis of interference and coverage in terahertz networks. Nano Communication Networks 13:9–19.

[33]   Petrov V, Komarov M, Moltchanov D, Jornet JM, and Koucheryavy Y (2017) Interference and SINR in millimeter wave and terahertz communication systems with blocking and directional antennas. IEEE Transactions on Wireless Communications 16(3):1791–1808.

[34] Pierobon M, Jornet JM, Akkari N, Almasri S, and Akyildiz IF (2014) A routing framework for energy harvesting wireless nanosensor networks in the terahertz band. Wireless Networks 20(5):1169–1183.

[35] Han C, Jornet JM, Fadel E, and Akyildiz IF (2013) A cross-layer communication module for the Internet of Things. Computer Networks 57(3):622–633.

[36] Elayan H, Johari P, Shubair RM, and Jornet JM (2017) Photothermal modeling and analysis of intra-body terahertz nanoscale communication. IEEE Transactions on Nanobioscience 16(8):755–763.

[37] Rahani EK, Kundu T, Wu Z, and Xin H (2011) Heat induced damage detection by terahertz (THz) radiation. Journal of Infrared, Millimeter, and Terahertz Waves 32(6):848–856.

[38] Akyildiz IF, and Jornet JM (2010) The internet of nano-things. IEEE Wireless Communications 17(6):58–63.

[39] Abadal S, Alarcón E, Cabellos-Aparicio A, Lemme M, and Nemirovsky M (2013) Graphene-enabled wireless communication for massive multicore architectures. IEEE Communications Magazine 51(11):137–143.

[40] Chen Y, and Han C (2018) Channel modeling and analysis for wireless networks-on-chip communications in the millimeter wave and terahertz bands. In: IEEE Conference on Computer Communications Workshops, pp. 651–656.

Chapter 3

Simulation and experimental platforms for nano-electromagnetic communication networks

Yu Zhou[1], Shaolong Shi[2], Junfeng Xiong[2], Yifan Chen[3,4], U. Kei Cheang[2] and Qingfeng Zhang[2]

3.1 Introduction

With the recent progress in semiconductor materials technology [1], engineering bacteria technology [2], and nanorobotic technology [3,4], especially in the field of medical applications, a new type of communication called molecular communication [5] has become a popular topic of research. Different from classical wireless communications which use electromagnetic waves as the information carriers, the carriers in molecular communication are molecules. Using magnetism-sensitive molecules as the information carriers allows for control through the use of an external electromagnetic field. Furthermore, observation and monitoring in real-time using existing electromagnetic imaging technologies become possible. The small-sized information carriers and the external electromagnetic manipulation feature make such a system a prime candidate for a wide range of applications in the fields of medicine, environment, etc., especially in the field of targeted drug delivery.

In traditional drug delivery systems, such as oral or intravascular injection, systemic blood circulation will carry the medicine constantly cycling around the entire body. In the process of cycling, most therapeutic agents cannot reach the organ to be treated but are absorbed by healthy tissues. Conventional chemotherapy shows that only a small fraction of the drug (0.1%) is taken up at the tumor cells, while the remaining part (99.9%) is absorbed by healthy tissues, which causes huge side effects on the human body. To address this problem, it is necessary to develop

[1]Beijing Institute of Collaborative Innovation, Beijing, China
[2]Department of Electrical and Electronic Engineering, Southern University of Science and Technology, Shenzhen, China
[3]Faculty of Life Sciences, University of Electronic Science and Technology of China, Chengdu, China
[4]School of Science and Engineering, University of Waikato, Hamilton, New Zealand

engineered systems to increase the targeting capabilities of drug carriers; one possible way is to combine biodegradable semiconductor materials and engineered bacteria with nanorobotic technologies to create inorganic or organic miniature robots that can actively target deceased area in the human body. Studies of engineered bacteria have verified the non-toxic feature in bioproducts and medical applications [6]. Here, nanorobots are defined as micro-entities that are manipulated using feedback control and can perform various functions using components less than 100 nm; the size of the nanorobots is consistent with the size of nanoparticles used for particulate drug delivery systems.

Due to the necessity of developing a computational framework to enhance the targeting capability of nanorobots, it is important to first discuss nano-electromagnetic communication networks by combining communication theory and medical applications together. In this chapter, these two will also be described as a single integrated concept. From the perspective of drug delivery systems, a cross-scale transient nanorobotic platform for transporting a pharmaceutical compound in the human body can be summarized as follows: under the navigation of an external control unit, miniature degradable robots will perform magnetic field sensing, guided directional swimming, as well as drug-carrying and releasing *in vivo* [7]. Drug particles are the cargo loaded onto nanorobots by utilizing biological or chemical bonding methods [8]. They can also be functionalized by means of proper molecular groups such as peptides or antibodies, which are able to bind to receptors at the unloading destination [8]. In order to deliver sufficient amounts of the drug to a destination, a swarm of nanorobots can be injected simultaneously and controlled like a unified organism [9]. The functions of these nanorobots may include field sensing, drug carrying, drug release, and propulsion. The nanorobots will be resorbed by the human body after completing the tasks. According to Chahibi *et al.* [10], the size of a single nanorobot (diameter of 1–2 μm) is smaller than half the diameter of the tiniest capillaries (~8 μm), which means that no obstructions will occur when nanorobots are transporting. From a communication perspective, the drug particles and nanorobots are message carriers. The vasculature is the propagation channel. The loading/injection and unloading of the drug particles are the transmitting and receiving processes. The most striking difference between such a system and a traditional wireless communication system is that the communication process can be controlled and tracked, which is similar to controlling through simple or multi-touch gestures by a finger through a touch screen.

A new field of communication naturally requires a new standard protocol to regulate. As such, a new standard working group, the IEEE 1906.1, is committed to establish a standard for nanoscale and molecular communications [11] and build a framework on the Network Simulation 3 (NS-3) platform in accordance with the specifications they have proposed with several example modules.

In this chapter, we will introduce a general system model for nano-electromagnetic communication networks based on the NS-3 framework and explain each element in the context of drug delivery application. Finally, we will also describe an experimental platform that can be used to validate the simulation model.

3.2 System architecture

This section will first describe the architecture of a cross-scale targeted drug delivery (TDD) system and give a comparison between TDD and classical wireless communication systems. Subsequently, the components and structure of the 1906.1 framework will be introduced.

3.2.1 Drug delivery system

The cross-scale drug delivery system includes an external macro-unit (MAU) and a number of *in vivo* drug-loaded micro-units (MIUs). As shown in Figure 3.1, the MAU directs the motion of a swarm of nanorobots by generating a guiding field [3,12]. The MAU also applies angiography to partially visualize the inside, or lumen, of blood vessels in the human body. For tracking of the swarm, drug particles could be labeled with fluorescent quantum dots. This procedure enables the drug cargo to feed the location of the swarm directly back to the MAU [13,14]. Thus, drug cargo in the form of a contrast agent can be employed. The MAU performs contrast imaging to localize the swarm [15].

This nanorobots-assisted drug delivery system can also be described using the nanoscale communication model: nanorobots and drug cargo are message carriers; blood vessels are the channel for information exchange; the loading and injection behaviors correspond to the transmitting process, while the unloading behavior corresponds to the receiving processes. Table 3.1 gives a comparison between the

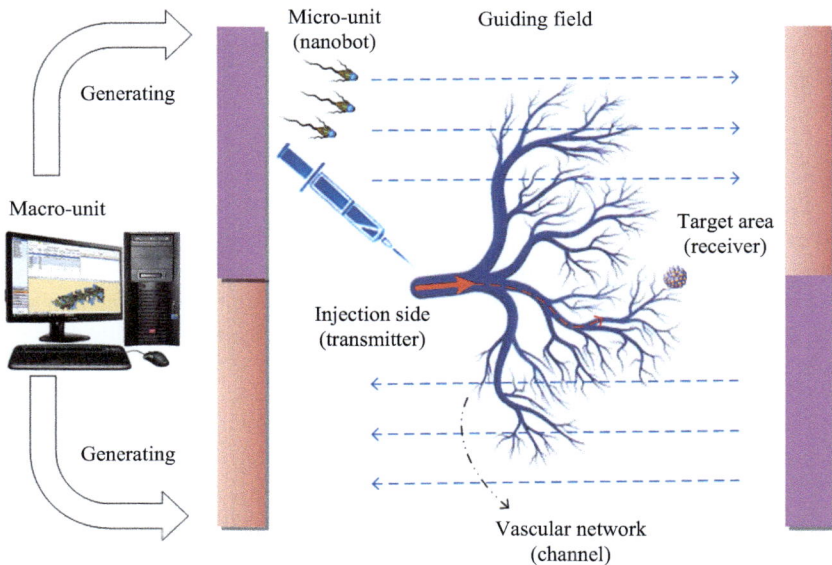

Figure 3.1   Conceptual illustration of the cross-scale TDD system

Table 3.1   Comparison between wireless communication systems and some nanocommunication systems

Components	Diffusion-based NanoCom	NanoCom in TDD	Wireless communication
Transmitter	Molecule	Injection	Antennas
Receiver	Molecule	Pathological tissue	Antennas
Channel	Aqueous medium	Vascular network	Air
Information	Chemical signal	Drug molecules	Analog and digital signals
Carrier	Molecule	Bacteria, magnetic-sensing nanomachine, etc.	Electromagnetic wave
Motion mechanism	Brownian motion	Diffusion, magnetic-control, optical-control, etc.	Radiation

wireless communication system and the nanocommunication system via diffusion and TDD application.

3.2.2   1906.1 framework

The 1906.1 framework defines a set of fundamental components in the nanoscale communication system as follows [11]:

Field: (A) It is a generic IEEE 1906.1 component that provides the service of controlled motion of message carriers. The application of the field component can result in message carriers whose motion has both a deterministic and a random component. (B) A region of space throughout which the force produced by an agent or agents, such as an electric charge, is operative

Medium: The interface connecting the transmitter and receiver, which can include gas, gel, or liquid.

Message carriers: The physical entities that convey the message across the medium.

Message: The information to be conveyed, which is known to the transmitting party and unknown, but recognizable, to the receiving party. The definition of a message includes signals transmitted for control purposes.

Motion: A generic IEEE 1906.1 component that provides the service of enabling a message carrier to move. Motion can be passive, in which message carriers are caused to move by forces in the medium or active, in which message carriers power their movement. See also, message carriers.

Perturbation: A generic IEEE 1906.1 component that provides the service of changing the affinity, motion, placement, or flow of message carriers in order to form a signal.

Specificity: (A) A generic IEEE 1906.1 component that provides the service of enabling a message carrier to convey its information to a desired receiver or a class of receivers while avoiding conveying information to other receivers or classes of receivers. (B) In biology, it is the ability of a protein binding site to bind to specific

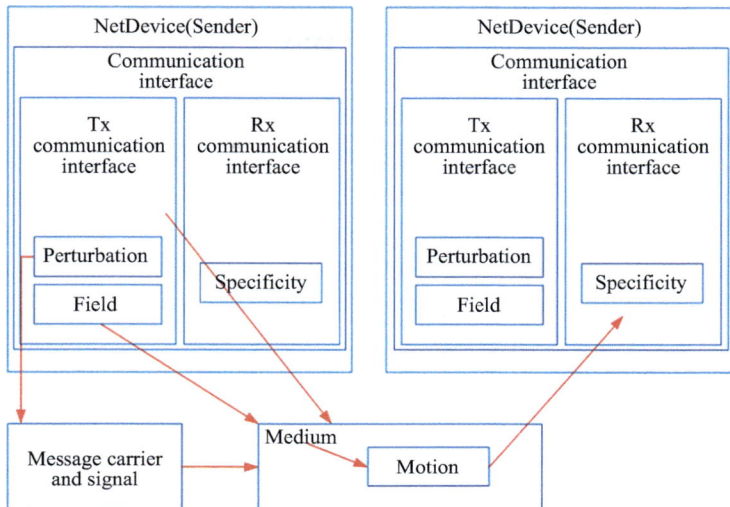

Figure 3.2   Reference model of 1906.1 framework [11]

ligands; the fewer ligands a protein can bind, the greater will be its specificity. Contrast: sensitivity.

In addition, each component shall have a general, but clearly defined interface to the components for which it provides. Examples include message-to-message-carrier (encoding), message-carrier-to-motion (range of motion), motion-to-message carrier (controlled motion), field-to-perturbation (rapid control of field), perturbation-to-specificity (ability to dynamically change specificity to encode a message), specificity-to-message-carrier (message carrier and binding capability), and message-carrier-to-receiver (decoding). Figure 3.2 indicates the relationships between all components and the corresponding interfaces inside the framework.

3.3   Simulation platform

In this section, a simulation platform for nanocommunication network based on the 1906.1 framework will be illustrated.

3.3.1   Message carrier

Message carriers in the nanocommunication network are externally controllable and trackable nanorobots which have a "cargo" to store the drug particles and achieve directional movement by sensing the external magnetic field. Thus, at least three basic modules are necessary to compose these nanorobots: navigation and sensing, propelling, and cargo. A more detailed structural model of the nanorobots is provided in [16], which also discusses the related potential technologies that might be applied to realize the functions. In this section, we will only focus on utilizing the key attributes to complete the network construction. The design of nanorobots will be introduced in the experiment platform section. In a targeted drug delivery system, the performance is highly related to the final concentration of drugs accumulated in the targeted area which is dependent on the concentration of nanorobots at the destination. This concentration is determined by the system architecture and the absorption efficiency associated with the drug resistance and sensitivity of the target cell. In the message carriers module, we only define these physical properties of the carrier that would affect the final concentration. The key parameter in this component is the lifespan of nanorobots, which represents the stable working period after injection. After evaluating the propagation time of nanorobots in the vascular channel, a reasonable lifespan could be selected to guarantee the performance of drug delivery.

3.3.2   Medium (channel)

As previously mentioned, the environment or the channel in a TDD system for carriers to move is the microvascular environment of the human body. Therefore, this chapter also takes establishing the model of microvascular environment as an example to illustrate the channel modeling method of a nanocommunication network.

Due to the limitations of modern medical imaging technologies, it is difficult to obtain the exact values of all of the important vascular parameters, such as topology, size, orientation, and speed of blood flow. Thus, it is useful to utilize statistical approaches to simulate the vascular morphology and the fractal model is chosen as the basic structure [16]. This is because the distributive function of vasculature is to bring a stream of blood to every living cell within the body or organ and is achieved by successive dichotomous division or bifurcation. The self-similarity of bifurcation at each node gives the vascular connection a fractal character. In the system model provided in [17], there are three consecutive components, which are a fractal-based angiographically resolvable region, a fractal-based angiographically unresolvable region, and a near-target region. The first two components are called the far region, which corresponds to the arteries and is modeled using a fractal-based bifurcating structure. On the other hand, the third component is called the near region, which corresponds to the capillaries and is simply modeled as a homogeneous aqueous medium without looking into the actual structure of microcirculation. The relations of these three regions are demonstrated in Figure 3.3.

The far region describes the movement of a nanorobot swarm in the angiographically visible blood vessels. Vasculature in the human body exhibits successive dichotomous division or bifurcation. The relationship between the diameter of the parent vessel and the diameters of two daughter vessels at each node follows the generalized Murray's law as demonstrated in Figure 3.2 [18,19]:

$$d_p^x = d_1^x + d_2^x \tag{3.1}$$

where d_p, d_1, and d_2 are the diameters of the parent segment, the first daughter segment, and the second daughter segment, respectively, as shown in Figure 3.4.

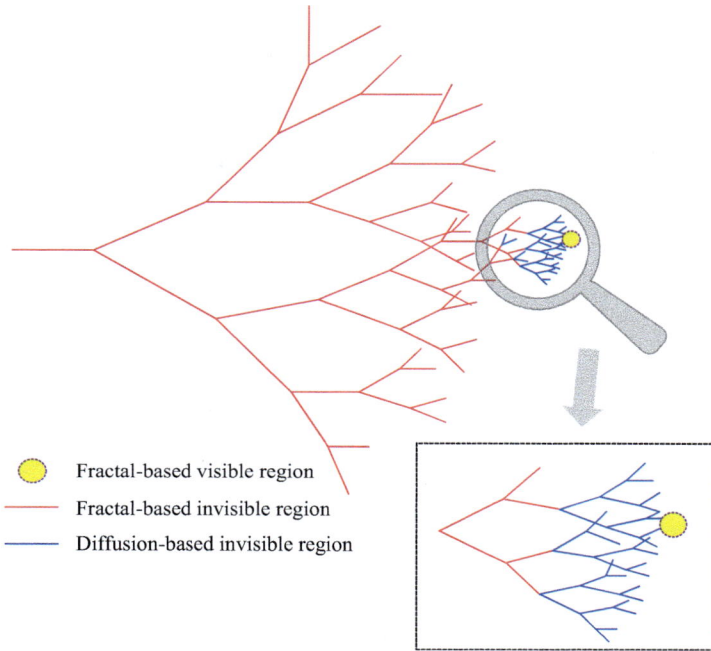

Figure 3.3 Relationship between the near region and the far region

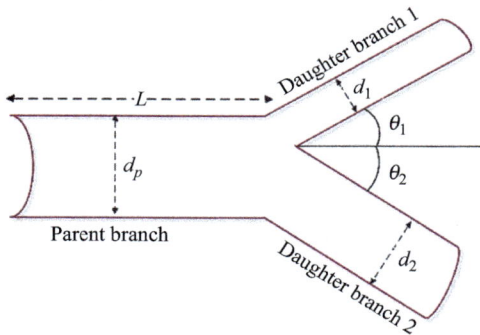

Figure 3.4 Bifurcation in the vascular system

The exponent x is called the bifurcation exponent, which ranges from 2.0 to 3.0 [19]. The vessel length is proportional to its diameter:

$$L = Kd \qquad (3.2)$$

where K is a constant for given vessels. For small arteries, $K \approx 60$ [18].

Vascular bifurcations often exhibit asymmetric structure such that the two branch diameters differ from one junction to the other. The ratio of the two diameters is defined as

$$\lambda = d_1/d_2, \quad d_1 \le d_2 \tag{3.3}$$

where arterial vessel λ is in the range of 0.59–0.83 [19]. In the following simulation, this parameter is set to be a random value varying between 0.59 and 0.83 for every bifurcation.

Murray's formulas also describe the angle between two daughter vessels, as shown in Figure 3.2. Define θ_1 and θ_2 to be the angles between daughter branch 1 and the parent segment, and daughter branch 2 and the parent. The following relationships can be obtained [19]:

$$\cos\theta_1 = \frac{d_p^4 + d_1^4 - d_2^4}{\left(2d_pd_1\right)^2}, \quad \cos\theta_2 = \frac{d_p^4 + d_2^4 - d_1^4}{\left(2d_pd_1\right)^2} \tag{3.4}$$

The total branch angle $\theta_1 + \theta_2$ between two daughter vessels after division is known as the bifurcation angle, which ranges between 75° and 90°. The arterial branching angle falls roughly in the range of 60°–80°. For capillaries, this angle is between 80° and 90° [19].

According to our simulation results, at every bifurcation where Murray's law applies, the diameter of the daughter branch is around 70% of that of the parent branch. Thus, to satisfy the variation of vessel diameter from the artery (~50 μm) to capillary (~8 μm), we need at least 7 consecutive bifurcations (i.e., 7 levels) from the injection site to the tumor microvasculature. The current simulation study considers 7–10 levels of bifurcations.

Corresponding to the far region, the near region describes the environment surrounding the target tissue or cells which connect to the distal ends of neighboring vascular trees. This particular area needs to be distinguished from the previous fractal network because the environment becomes very different: firstly, the blood vessels are invisible in angiography since the diameters are fairly small; secondly, all capillaries are interconnected rather than being fully separated [20], thereby forming a complex fluid network around the destination. Currently, there is no more accurate and comprehensive model to simulate this region. In this section, we provide some feasible solutions approximated from one or some aspects, such as the diffusion model, grid model, and invasion percolation model. Since these models are still in the research stage, this section will only present a preliminary introduction.

The diffusion model replaces the complex interconnected vascular network surrounding the receiver with a fuzzy area which comprises two circular layers, as shown in Figure 3.5(a). The inner layer represents the destination—tumor cells, which is surrounded by an outer ring connected to multiple vascular trees. The radius of the inner layer is around tens of micrometers, comparable to the highest resolution achieved by the state-of-the-art imaging techniques [21]. The outer ring

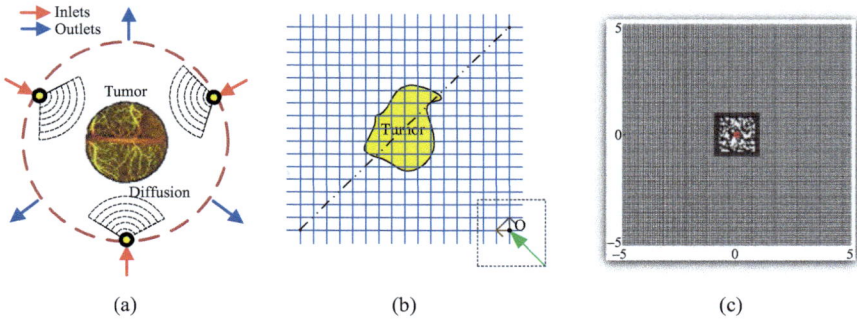

Figure 3.5   (a) Diffusion model; (b) grid model; and (c) invasion percolation model

defines the interface between the far region and the inner layer. Nanorobots may enter the near region from different vascular branches on the same tree or from different trees.

The grid-like model assumes that the normal tissue is regularly vascularized, which results in a latticed capillary network around the target, as shown in Figure 3.5(b) [17]. The basis for this assumption is typical skeletonized images of various classes of vascular networks and demonstrates that normal capillaries network always experience uniformly distributed two-dimensional patterns to guarantee the sufficient oxygen supply toward the tissue. Thus, it is a common practice to restrict the vessels to run only parallel to the coordinate axes [22]. The capillaries crisscross together around the target area and the trajectory of nanorobots is mainly affected by the direction of blood pressure.

The invasion percolation model has some similarities with the grid model but is more complex. It also restricts the vessels to run only parallel to the coordinate axes and establish a lattice network. The normal tissue is assumed to be regularly vascularized, which results in a uniform capillary network, where a given microvascular density is determined by the distance between the capillaries, as shown in Figure 3.5(c) [23]. The closer to the tumor, the smaller the spacing of the blood vessels and the higher the density of the network. The fractal dimensions of tumor vasculature indicate that invasion percolation, which is a statistical growth model adjusted by local substrate properties, is a superior illustration for the tortuous vessels and extensive avascular space exists in tumor. The potential paths of vascular growth are represented by some discrete points in a square lattice structure in this computational model. By allocating a uniformly distributed intensity random value to each point on the underlying grid, an initial state of invasion percolation is obtained. Afterward, from any arbitrary site, the network expands to the grid point with the lowest intensity adjacent to the current position. This expansion process will iteratively repeat until the desired lattice occupancy is achieved. Assuming that all adjacent occupying grid points are connected by the blood vessels, blood fills the network from the starting entry point and withdraws from the prescribed exit point. The network is then trimmed to only retain the portions with non-zero blood flow, leaving the backbone of the infiltrated cluster. By considering the fractal dimension, the simulated network can be matched to the tumor vasculature in the real case, which can be achieved by selecting the appropriate occupancy level.

3.3.3   Field

When the nanorobot is injected into the human blood circulation system, how to guide the nanorobot to move in the complex vascular network is the core of the whole system. Researchers have found solutions to the problems of navigation systems both internally and externally.

In internal navigation solution, the nanorobot is integrated with its own sensor. Equipped with chemical or spectral nanosensors, it is possible to detect and find the right location based on specific chemical or optical tracking techniques.

This section is focused on external solutions, which use chemical signals, optical signals, ultrasonic signals, electromagnetic field, ++X-rays, or magnetic field to guide the movement of nanorobots. Major advances have been achieved to control the motion of nanomachines to steer them toward desired sites by inducing external gradients. These stimuli are categorized as follows.

Chemical stimuli can trigger swarming behavior of different types of catalytic nanomachines ranging from simple to more complex units such as Janus spheres and Au nanowires or nanoparticles. This method is highly dependent on the surface properties because it is based on the reaction on the surface. For instance, chemical-induced aggregate behavior that has been observed in self-propelled Au nanoparticles is the diffusion of ionic products from the catalytic Au surface. It contains chemical reaction between hydrazine and H_2O_2 catalyzed by the gold surface that generates chemical gradients on the nearby Au nanoparticles, which pull the particles toward the largest electrolyte gradient existed site [24]. However, for chemically triggered swarming behavior, the swarming time is highly dependent on the reaction speed, which limits its applications.

Optical stimuli have also been an efficient method for inducing aggregation of nanomachines. This mechanism is defined as the use of intensely focused light beams to accurately trap and manipulate nanomachines. This is the most convenient approach as it does not require complex processes for device fabrication. Intensity gradients that are generated by a converging beam are able to polarize nanomachines to move toward the highest gradient region of the electrical field. Light-driven nanoscale plasmonic motors have been proposed, which demonstrate the ability to control rotation and direction of different plasmonic modes by varying the wavelength of the incident light [24].

Another effective way to regulate the collective behavior of nanomachines is ultrasound, which is an on-demand motion control, a long lifetime, and a noninvasive aggregation method. Interactions between ultrasound and nanomachines lead to swarming phenomena based on pressure nodes or antinodes, where the nanomachines migrate from peaks and accumulate in the troughs of the associated standing waves [25].

Electrically triggered collective pattern is another collective method, which induces dielectrophoresis—a phenomenon exerted on dielectric nanomachines suspended in water solution to initiate the gathering of nanomachines. The application of this method for *in vivo* TDD has been limited because it requires complex configuration to generate an appropriate field and it requires high voltages that may damage tissue. Manipulation of nanowires suspended in a liquid by combined DC with AC electric fields was reported, which has been used to transport drug-carrying functionalized nanowires in a controlled manner to the target areas [24].

Magnetic stimuli are the most common way to remotely control the motion of nanomachines. Magnetic fields can be used to manipulate the motion of nanomachines as well as track their movement. This stimulus also has the capability of superfast operation as its response time is just a few seconds. Injected nanomachines loaded with drug molecules can be aggregated and guided by a local magnetic field, which can be applied in nearly all human tissues without any toxicity [24]. Table 3.2 gives a comparison between several different types of nanomachines and their control mechanisms.

A magnetic field controls the most widely used external guidance scheme at this stage. This chapter will also use this as an example for analysis. As the ability of magnetic sensing is integrated into nanorobots, an external magnetic field is introduced to guide the movement of nanorobots in the vascular network. Because the path loss (i.e., reduction in the nanorobots concentration) during transmission is closely related to the controlling mechanism, the field component will not only define the field but also cover the effects of the field in three different transmission environments as mentioned in Medium sections.

The fractal-based visible region represents the several stages of the vascular tree near the front end, that is, close to the artery, wherein the movement of the nanorobot group and blood vessels is angiographically visible. Due to the excellent visibility, the control process in this area is very precise, which will ensure that the nanorobot group moves along the correct path and greatly reduces losses.

On the other hand, the invisible area based on the fractal occupies several levels in the fractal blood vessel tree close to the target area. As the radius of the blood vessels continues to shrink with the extension of the network, it is almost impossible to clearly identify the structure of the blood vessels using existing angiographic techniques at the levels close to the target area, i.e., angiographically invisible. In this case, the guiding field cannot provide accurate guidance for the movement of the nanorobot swarm but only provides a gradient from the current

Table 3.2   Several types of nanomachines and corresponding control mechanisms

Type	Controlling mechanism	Propulsion
Magnetotactic bacteria	Magnetic field	Flagella
Magnetic nanoparticles	Magnetic field	Spin
Plasmonic motor	Light	Light-induced rotation
Chemotaxis nanomotor	Chemical gradient	Chemical reaction

position to the destination. The corresponding parameters include the strength of the field and the angle between the field and the current branch.

Regarding the use of magnetic fields in the near-field region, since there is no recognized accurate model at this stage, we can only briefly describe them based on the examples of several approximate models listed in the channel section. In the diffusion model, the nanorobot is diffused in a relatively open environment from each entrance. As the uncertainty of free diffusion will greatly reduce the transportation efficiency, the diffusion model utilizes a magnetic field to create a series of aggregation points along the direction from the initial entrance toward the target, allowing the nanorobot to undergo a repeated process like diffusion-aggregation-diffusion-aggregation [17]. In the lattice network model, the nanorobot moves from the hypotensive side of the network to the hypertensive side under the influence of blood pressure. Once an offset angle between the center of the target area and the blood pressure direction exists, the diffusion efficiency of the nanorobot will reduce. The role of the magnetic field is to correct or to compensate for the existence of the offset [26]. In the invasion percolation model, the magnetic field no longer exists as a guiding factor, but only to drive the movement of the magnetic nanorobot. The guiding issue will be accomplished by the nanorobots swarm that explore the optimal path with evolutionary algorithms [23].

For specific parameters and control mechanisms of the magnetic field, different types of nanorobots may correspond to different control structures. For instance, micro-electromagnets [27] or electrical conductor networks [4] could be used to generate the controlling field for nanorobots using magnetotactic bacteria (MTB) as a carrier. For nanorobots composed of nanomagnetic particles, a controllable magnetic field can be generated by a cube-like 3D coil system [4]. Under the control of corresponding fields, different types of nanorobots will also show different characteristics. Engineered bacteria type could achieve a high motion velocity (close to the velocity of blood flowing in capillaries) and is easy to be guided in a large swarm. In contrast, the artificial magnetic particle type could be flexibly assembled together or disassembled into small units and complete the very precise movement. In our work, we will focus on whether these controlling methods could provide directional guidance to nanorobots, and the velocity of nanorobots could be achieved under control. This velocity is a key parameter involved in the calculation of the next component (i.e., motion component). To explain the motion mechanism, some detailed information about how the magnetic field guides the movement of nanorobots inside the vessels is introduced in the next section, along with the motion calculation.

3.3.4   Motion

This section will use the motion mechanism of nanorobot in the fractal-based network as an example to explain how to establish a simulation model. This example is only focused on the movement of nanorobots in the *far region* but not in the *near region*. The near region movement highly depends on the model applied and owns a huge difference from each other.

3.3.4.1 Motion component in fractal-based visible region

In order to analyze the movement of nanorobots in a vascular network, the first issue that needs to be addressed is how to describe blood flow. Due to the complexity of the human vascular network, it is difficult to obtain a precise description of the blood flow velocity in different environments. An approximate solution is to treat the human circulatory system as a set of complex enclosed pipelines in which the liquid flowing is the blood flow. Due to the smoothness of blood vessels, blood flow may present either a turbulent (i.e., chaotic) or a laminar (i.e., smooth) feature. In hemodynamics, Reynold's number (denoted *Re*) is a dimensionless relationship that determines the behavior of fluid in a tube, in this case, blood in the vessel. The equation for this relationship is written as

$$Re = \frac{\rho v L}{\mu} \tag{3.5}$$

where ρ is the density of blood, v is the mean velocity of blood, L is the characteristic dimension of the vessel (in this case the diameter), and μ is the blood viscosity.

Once the type of liquid flowing in the pipeline is fixed, this dimensionless parameter is proportional to the diameter of the pipeline and the internal liquid velocity. When Reynold's value is less than 2,300, all liquids exhibit laminar motion, which is characterized by a constant flow motion. Turbulence only dominates flow state while the Reynolds number exceeds 4,000.

The most significant parameters for the TDD system applies to the capillary part of the vascular network. Since the blood velocity, vessel diameter, and the blood viscosity constantly fluctuate with the shape of the blood vessel, the average values or approximation values are applied for these parameters during calculation: $v \approx 0.3$ mm/s [28], $\mu \approx 3 \times 10^{-3}$ Pa s [29], $L \approx 0.6$ mm, $\rho = 1.06 \times 10^3$ kg/m^3.

The Reynolds number at the capillaries is around 7×10^{-5}, which is several orders of magnitude smaller than the threshold where turbulence occurs, due to the relatively small radius and velocity. Therefore, the blood flow in the capillaries is assumed to be laminar. Similar situations also occur in the other parts of the vascular network where blood velocities are low, such as veins and arterioles. The only exception happens in the aorta near the heart, which experiences a large velocity. Blood flow in this part mainly presents in a turbulent state [30].

When the laminar flow becomes the primary state of liquids, Poiseuille's law, which indicates that the flow rate of viscous fluid is proportional to the biquadrate of the diameter of the rigid pipe, is introduced to analyze the circulatory system and explore blood flow characteristics [31,32]:

$$Q = \frac{\pi R^4}{8\mu} \frac{\Delta P}{L} \tag{3.6}$$

where Q is the blood flow volume, μ is the viscosity of the blood, ΔP is the difference of blood pressure at the two ends, and R and L are the radius and length of the vascular pipe, respectively.

In this case, the viscous force between the blood and the vessel wall gives blood velocity a parabolic profile, where the largest velocity occurs at the center of the vessel and the lowest value near the vessel wall, as shown in Figure 3.5. The velocity distribution along the blood vessel can be derived from Poiseuille's law [24]:

$$v_x(t) = \frac{1}{4\mu} \frac{\Delta P}{L} \left[R^2 - r^2(t) \right] \tag{3.7}$$

where $r(t)$ is the distance between the swarm and the centerline of the vessel at propagation time t.

Assuming that the size of the swarm is much smaller than the diameter of a large blood vessel, the entire swarm can be considered as a point in the blood vessel. Then, the motion of the nanorobot swarm between the two branches can be theoretically categorized into two different scenarios based on (3.5)–(3.7), as shown in Figure 3.6.

Lateral movement (LM) from A to B
In this scenario, the swarm swims along the same side of the vessel. Only the horizontal component of the velocity v_x contributes to the propagation delay. For LM, $v_x = v_x(0)$ remains unchanged until the swarm enters the next branch. The total propagation delay from A to B is obtained by applying the following equation:

$$t_p = \frac{L}{v_x(0)} = \frac{4\mu L^2}{\Delta P[R^2 - r^2(0)]} \tag{3.8}$$

Diagonal movement from A to C
In this case, the nanorobot completes a diagonal direction of movement, that is, across the parent branch and enters the target daughter branch on the opposite side. Assuming that the external guiding field drives the nanorobot to generate a constant component v_y in the vertical direction, then v_y should be sufficiently large to ensure that the swarm passes through the blood vessel before reaching the next bifurcation.

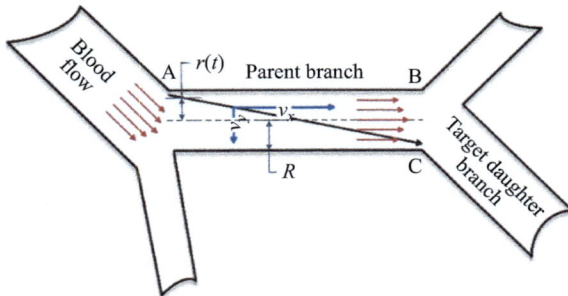

Figure 3.6   *Movement and velocity of the nanorobot swarm between two bifurcations*

The maximum moving distance of the nanorobot in the vertical direction is required to reach $\gamma \times 100\%$ of the diameter of the parent branch, where γ depends on the diameter of the target daughter branch as compared to the parent branch. According to the simulations of vasculature governed by Murray's law, γ is taken to be 70%. With these constraints, the propagation delay for the diagonal movement (DM) can be obtained as the solution to the following equation:

$$\int_0^{t_p} \frac{1}{4\mu} \frac{\Delta P}{L} [R^2 - (R - v_y t)^2] dt = L \tag{3.9}$$

with $v_y t_p \geq \gamma d_p$. The propagation delay in the DM case can be ignored because of the tiny vessels at the distal end of the vascular tree. In terms of path loss, the DM case will be discussed in the following section.

3.3.4.2 Path loss in fractal-based invisible region

In clinical treatment, the concentration of the drug is strictly required to be kept in a specific range, which is called a therapeutic window or a safety window [33]. Therefore, the amount of drug particles successfully reaching the targeted site in the TDD system is the most critical performance measure, which corresponds to the classical parameter of path loss in the context of the nanocommunication network. Due to the imperfection of the near region model, only the path loss of the far region will be discussed, namely, the fractal-based regions. The path loss in the far region is given by the percentage of nanorobots successfully delivered to the near region, which may be affected by various attenuation mechanisms, including diffusion, branching, degeneration, and advection. Different loss factors occupy different positions in different systems. Loss factors such as diffusion and advection, due to continuous manipulation and navigation, account for a low proportion of total path loss in targeted drug delivery systems, especially in the fractal structure model. However, in some nanocommunication networks using diffusion as the main mechanism, these factors may be dominant. The degradation is highly related to the material used to manufacture nanorobots, which are often simplified by a scaling factor during the modeling process. Once this scaling factor is set to 0%, it assumes that the dissolving speed of biodegradable material (the lifespan) is much larger than the transmission delay, and thus no degeneration loss exists. In this chapter, the branch loss (i.e., the percentage of nanorobots entering the wrong branch as they pass each fork) will be the most important factor in the path loss of the targeted drug delivery system in the following discussion.

Although the model mentioned above contains two fractal-based regions, the branch loss will only occur in the deeper layers of the vascular tree, that is, the angiographically invisible region. In the visible region, a clear image of the blood vessel ensures precise control of the nanorobot swarm, while the larger diameter and length of the blood vessel also provide sufficient space for the nanorobot to find the correct path. In the invisible region, the guiding field only provides a gradient to the destination, as shown in Figure 3.7(a). Obviously, the branch loss is closely related to the above DM scenario, and the angle between the guiding field and the parent branch plays an important role in its calculation.

The nanorobot swarm can be simplified to a point to analyze its motion when the vessel size is large. This assumption is no longer applicable as the diameter of the deep blood vessels becomes narrower and closer to the size of the nanorobot swarm. Therefore, when considering branching loss, the nanorobot is assumed to be evenly distributed along the cross-section of the blood vessel. The exact value of the branch loss is determined by the proportion of nanorobots entering the two daughter branches. As shown in Figure 3.7(a), the velocity component in the vertical direction of the parent branch determines whether the nanorobot can successfully reach the correct daughter branch. Suppose that the lower branch is the right path to the destination, the nanorobot must reach the shaded area at the bifurcation to enter that path, as shown in Figure 3.7(b). Subsequently, the percentage of nanorobots accumulated in this area is expressed as

$$p = (1 - \gamma) + \frac{a \sin \theta v t}{\gamma d_p} \tag{3.10}$$

where θ is the angle between the field and the parent branch, t is the propagation duration of nanorobots in the parent branch, γ and d_p are the same parameters used in the DM scenario, which represent the diameter of the targeted daughter branch as compared to that of the parent branch and the diameter of the parent branch, v is the velocity of nanorobots driven by the guiding field, which is equal to v_y in (3.9), and a is an enhancement factor related to the strength of the guidance field. Note that the reason for setting the enhancement factor is that it is difficult to pinpoint precisely the bifurcation connecting the parent branch to the daughter branches due to resolution limitations. The enhancement factor describes the compensation effect of increasing the strength of the external control field to accelerate the diagonal motion to ensure the nanorobot swarm enters the correct path.

The variation of angle θ makes the percentage in (3.9) fluctuated, which leads to the variation in the vertical velocity. If the transmission of nanorobots along the

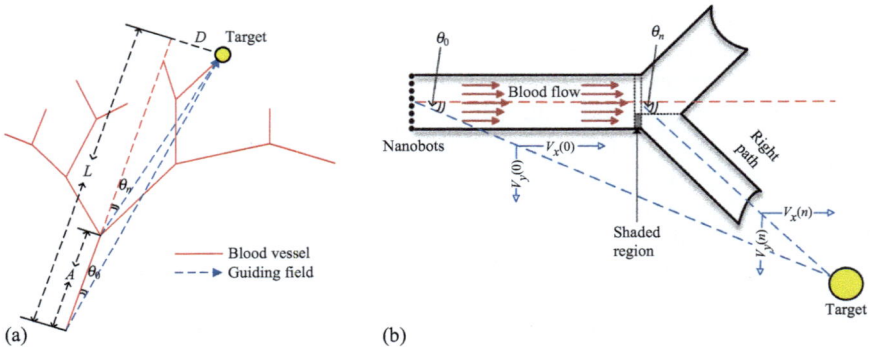

Figure 3.7   (a) Geometric relationship between the guiding field and the blood vessels and (b) the zoomed-up sketch depicting the movement of nanorobots for calculating the branching loss in the far region

parent branch is divided into multiple time slots, p_i will represent the percentage of nanorobots arrived at the shaded region at the ith time slot. As the velocity of nanorobots keeps changing, a more accurate measure is obtained from the average velocity, as shown in Figure 3.7(b):

$$p_n = (1 - \gamma) \pm \frac{at \frac{\int_{\theta_0}^{\theta_n} \sin \theta_i v d\theta_i}{\theta_n - \theta_0}}{\gamma d_p} \tag{3.11}$$

where θ_i $(i = 0, 1, \ldots, n)$ denotes the angle between the field and the parent branch at the ith time slot. Equation (3.10) can be simplified as

$$p_n = (1 - \gamma) \pm \frac{avt_n}{\gamma d_p} \times \frac{\cos \theta_n - \cos \theta_0}{\theta_n - \theta_0} \tag{3.12}$$

Following from Figure 3.6(a), the angle θ_i is calculated as $\theta_i = \tan^{-1} D^2/(A^2 - (v_L t_i)^2)$, where v_L is the sum vector of the blood flow velocity v_B and the horizontal component v_x due to the guiding field. As $v_x \ll v_B$, $v_L \approx v_B$ and $v_L t_i = 0$ when nanorobots have just entered the parent segment and $v_L t_i = L$ when nanorobots are at the bifurcation. Subsequently, for $\gamma = 0.7$ and $a = 1$, (3.11) reduces to

$$p_n = 0.3 \pm \frac{\cos\left(\tan^{-1}\frac{D^2}{A^2 - L^2}\right) - \cos\left(\tan^{-1}\frac{D^2}{A^2}\right)}{\tan^{-1}\frac{D^2}{A^2 - L^2} - \tan^{-1}\frac{D^2}{A^2}} \tag{3.13}$$

Equation (3.13) is the final expression of branch loss in the fractal-based invisible region.

3.3.5   *Tracking*

In the classical wireless communication network, there is no such module for tracking due to the invisibility and uncontrollability of electromagnetic waves. However, in nanocommunication networks, nanorobots as information carriers are generally visible and controllable. In order to achieve more effective control, observation, and tracking of nanorobots and their motion environment is an indispensable means in most cases. Although this module does not have any meaning for all nanocommunication networks, it is still worth discussing separately here.

In nanocommunication networks, the working environment of nanorobots such as human blood vessels is usually the micro-environment with the characteristics of small-scale and non-line-of-sight. Therefore, tracking the movement of nanorobots requires the support of medical imaging technologies. The most commonly used technique for high-resolution medical imaging is magnetic resonance imaging (MRI). To perform a study, the person is positioned within an MRI scanner that forms a strong magnetic field around the area to be imaged. In most medical applications, protons (hydrogen atoms) in tissues containing water molecules create a signal that is processed to form an image of the body. First, energy from an oscillating magnetic field is temporarily applied to the patient at an appropriate resonance frequency. The excited hydrogen atoms emit a radio frequency signal, which is measured by a receiving coil. The radio signal may be made to encode position information by varying the main magnetic field using gradient coils. As these coils are rapidly switched on and off, they create the characteristic repetitive noise of an MRI scan. The contrast between different tissues is determined by the rate at which the excited atoms return to the equilibrium state. Exogenous contrast agents may be given to the person to make the image clearer [33].

The highest resolution that MRI technology can achieve now is about 0.1–1 mm. This technology is widely used in clinical practice. In recent years, several research groups adapted the use of MRI to observe nanorobots *in vivo* [33]. However, the shortcomings of MRI are also obvious, including the high cost of equipment, slow imaging speed, and the inability to provide three-dimensional images.

Another option for tracking is using microwave imaging technology. According to the frequency and wavelength applied, common microwave imaging technologies include millimeter-wave imaging, sub terahertz imaging, and terahertz imaging. The principle is to irradiate the measured object with microwave and then reconstruct the shape or dielectric constant distribution of the object by the measured value of the external scattering field of the object. Since the dielectric constant is closely related to the water content of biological tissues, microwave imaging is very suitable for imaging biological tissues. When large discontinuities limit the efficiency of ultrasound imaging and the low density of biological tissue limits the use of X-rays, microwave imaging can play a unique role in obtaining information that other imaging methods cannot obtain. Compared with MRI, this technology, especially terahertz imaging, is safe, low-radiation, economical, and fast. In the field of medical diagnosis, compared to other frequencies, terahertz imaging has a shorter wavelength, less scattering in the human body, and better contrast ratio on cells or tissues such as tumors, fats, and glands.

Microwave imaging is an inverse scattering problem that inverts the extracted target feature information from the scattered echo signals. It relies on the interaction of electromagnetic waves with targets, mining and extracting target information from scattered echo signals, and reconstructing target features.

The main difficulty is that the microwave wavelength is close to the size of the measured organism, which causes the diffraction effect. The inverse scattering-based inversion algorithm for microwave imaging is more complex than the projection imaging method for X-ray.

3.4   Experiment

This section will take the magnetic-field-controlled nanorobot experimental control platform as an example to describe how to build an experimental platform that can be used to verify the simulation system model.

3.4.1   *Magnetic-driven nanorobot*

As described in section 3.3.3, there are many ways to externally control the movement of nanorobots, such as sound field drive, light drive, and magnetic field drive. Among them, the magnetic field drive is an effective and promising method for biomedical applications of nanorobots due to its low strength, low frequency, ability to penetrate biological tissues and be harmless to organisms. Different types of magnetic fields can be obtained by designing permanent magnets and electromagnetic coils with different configurations. By controlling the intensity, frequency, phase, and switching mode of the electromagnetic coil current in real-time, it is possible to provide power and controllability to the motion of magnetic nanorobots.

As long as the micro-nanorobots are given magnetic properties, the magnetic fields can be used to drive and control their motion. The application of magnetic fields in micro-nanorobots can be divided into two categories: (1) magnetic field-driven micro-nanorobots for which magnetic fields are used to provide actuation energy and to control movement direction and (2) magnetically oriented micro-nanorobots for which magnetic field is combined with other driving methods, where the magnetic field is only used to control the direction of motion.

A magnetic nanorobot can be driven by a magnetic field because it is subjected to the force generated by the gradient of the magnetic field or the torque due to misalignment to the magnetic field. The magnetic force and magnetic torque exerted on a magnetic object with a magnetic moment of m can be calculated by the following equations:

$$F(P) = (m * \nabla)B(P) \tag{3.14}$$

$$T(P) = m \times B(P) \tag{3.15}$$

where $F(P)$ and $T(P)$ refer to the magnetic field gradient force and the magnetic field torque of the nanorobots at point P in the magnetic field, respectively, and $B(P)$ refers to the magnetic flux density at this point.

In a uniform magnetic field, no magnetic force will be exerted on the nanorobots because the magnetic gradient does not exist. If the magnetic moment of the magnetic object is in the same direction as the uniform magnetic field, the magnetic torque will also be zero. Nanorobots will be subjected to the magnetic moment and deflected toward the direction of the applied magnetic field only when the magnetic dipole moment and magnetic field have different directions.

When the two directions are collinear, nanorobots will stop moving. Therefore, the uniform magnetic field cannot drive micro-nanorobots to achieve continuous movement. A variation of the magnetic field in time or space dimension is necessary. Figure 3.8 shows several magnetic fields currently used to drive nanorobots, including a uniform magnetic field rotating around the central axis [Figure 3.8(a) and (b)], a uniform magnetic field that oscillates a specific angle [Figure 3.8(c)], a pulsed magnetic field that intermittently produces magnetic induction [Figure 3.8(d)],

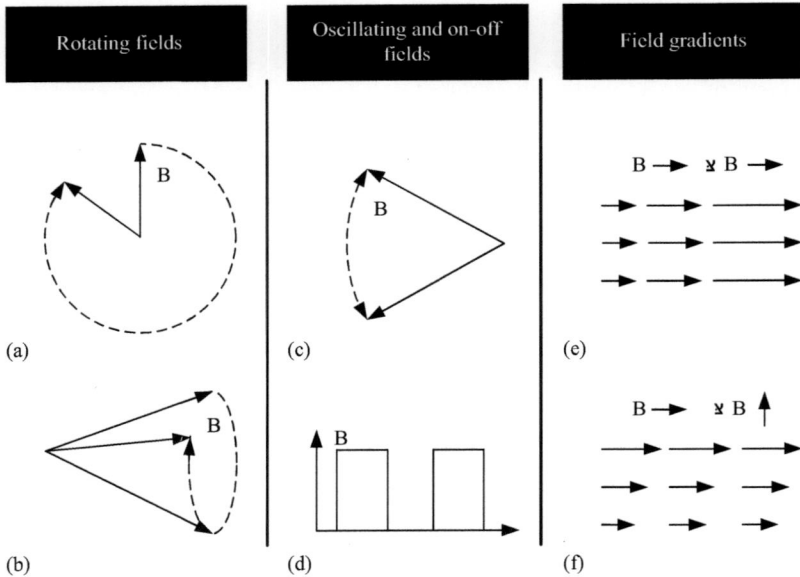

Figure 3.8   *The magnetic fields currently used to drive micro-nanorobots are:*
(a) Planar rotating magnetic field; (b) conical rotating magnetic field;
(c) oscillating magnetic field; (d) pulsed magnetic field; (e) gradient
magnetic field along the direction of the magnetic field; and (f)
gradient magnetic field vertical to the direction of the magnetic field
[34]. Copyright © 2013 Royal Society of Chemistry

and a magnetic field with a magnetic induction gradient in one direction
[Figure 3.8(e) and (f)] [34]. The gradient field belongs to a space-varying magnetic
field that varies with space, and any magnetically sensitive nanorobots can
response to the magnetic field force. Rotating, oscillating, and pulsed fields are
time-varying.

According to the aforementioned different ways of controlling the magnetic
field, most of the magnetic field-driven magnetic nanorobots commonly used in the
current research are divided into three types: (1) a spiral-driven micro-nanorobot;
(2) an oscillating magnetic field-driven flexible micro-nanorobot; and (3) a gradient magnetic field-driven micro-nanorobot and interface rolling micro-nanorobot.

3.4.2   Magnetic-guided nanorobot

The magnetic field, due to the aforementioned advantages, has become the dominant method for guiding the motion of micro- and nanorobots. Two widely used
nanorobots relevant to this work are magnetic-guided chemical-driven nanorobots
and magnetotactic bacteria.

Chemical-driven nanorobots are self-driven robots that convert chemical energy into kinetic energy. Such robots can react chemically to their environment and generate behaviors, such as self-electrophoresis, self-diffusion, and unidirectional jetting bubbles to promote their own motion. However, their direction of motion cannot be controlled by the chemical method, which inspires researchers to achieve steering control by introducing magnetic fields. For instance, Kline *et al.* [35] present a Pt/Ni/Au/Ni/Au multi-section nanorod; the Pt end is used for catalytic decomposition of H_2O_2 to generate self-electrophoresis phenomenon while the Ni end, pre-magnetized with a dipole perpendicular to the central axis of the body, is used for magnetic control.

Another special magnetically guided nanorobot is essentially different from the previous artificial magnetic particles. The nanorobots are living microorganisms widely distributed in the natural marine and lake environment called MTB. The currently known MTB are mainly aquaspirillum and bilophococcus. These bacteria have the ability to orient themselves along the magnetic field lines of the Earth with the help of a kind of self-generated nanoparticles called magnetosomes [3]. The magnetosome has a small volume (20–100 nm) with a shape of a truncated octahedron, parallelepiped, or hexagonal cylinder. The main components of a magnetosome are usually Fe_3O_4 and Fe_3S_4, which do not cause any toxicity to the human body. Normally, there are 2–10 magnetosomes existing inside each bacterium to achieve the guiding function, collaborating with the flagella to ensure the bacteria move in the correct direction. Figure 3.9 is a scanning electron microscope image of an MTB in which the distribution of magnetosomes can be clearly seen.

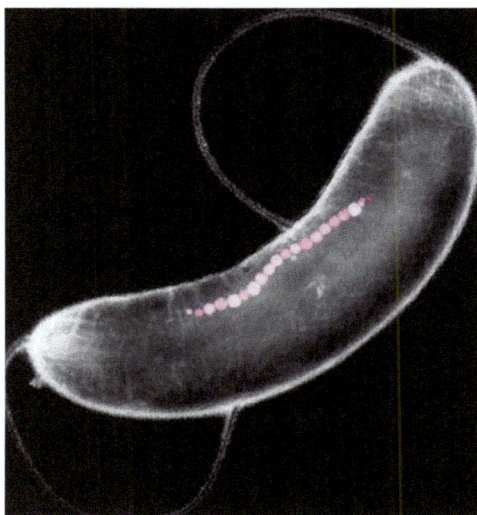

Figure 3.9   Magnetic-guided magnetotactic bacteria

3.4.3   Platform

In this section, we will introduce an experiment platform which consists of the nanorobots and the control system used to generate magnetic fields to drive the nanorobots. The system consists of three pairs of electromagnetic coils arranged in an approximate Helmholtz configuration, three power supplies (Kepco), a National Instrument data acquisition (DAQ) controller, a computer, an inverted microscope (Leica DM IRB), and a camera (Point Grey). The power supplies are programmable and can generate sinusoidal outputs to the coils to create a rotating magnetic field through the use of a DAQ controller. The camera provides visual feedback and records videos at 30 frames per second with a resolution of 1,280 by 1,024. The computer is used as a control interface for the camera and the DAQ controller [4].

The approximate Helmholtz coil system is designed to exert magnetic torque on the magnetic nanorobot without introducing translational force by generating a near-uniform magnetic field. The coils are arranged in a slightly different configuration than that of the normal Helmholtz coil, and thus we use the prefix approximate. Conventionally, Helmholtz coil restricts the distance between two coils of the same size to be the radius of the coils. Given the space constraint of the microscope, the configuration was designed to optimize the magnetic field profile in order to create a near-constant region at the center of the coils. In this study, the distance between the coils is equal to the outer diameter of the coils plus the thickness of the coils, creating a cube-like configuration for the 3D coil system (Figure 3.10). While the approximate Helmholtz coils cannot reproduce the same field due to the difference in coil distances, they are sufficient to generate a near-uniform magnetic field.

A type of nanorobot tested in the system consists of three firmly connected ferromagnetic beads (4.35 μm in diameter) forming an achiral structure with two planes of symmetry. The beads are linked using avidin–biotin chemistry to ensure structural integrity. Avidin-coated and biotin-coated beads are commercially available and will form chains when magnetized due to dipole to dipole attraction. Figure 3.11 illustrates the fabrication process and structure of the nanorobot.

The chirality of the microswimmer can be defined by its two planes of symmetry. According to Happel and Brenner, an object with two symmetry planes can produce forward propulsion while undergoing a rotation. The achiral magnetic particles are rigid due to the avidin–biotin linkage. A rigidity test was performed by quantitatively observing the relative distances (x, y, z) between the three firmly connected beads during swimming. The relative distances were obtained by performing 3D image processing on recorded videos of magnetic particles swimming. For each cycle of rotation, one frame was used where all three beads are visible. For consistency, the frame with which the magnetic particles show the same posture was used for every cycle. The x and y positions of each bead were obtained by computing the centroid of the beads using a circular Hough transform-based circle detection algorithm. The z position was approximated using a 3D tracking algorithm which worked based on the standard deviation of the intensity of the beads. First, a calibration curve was obtained to map the standard deviation of a bead's

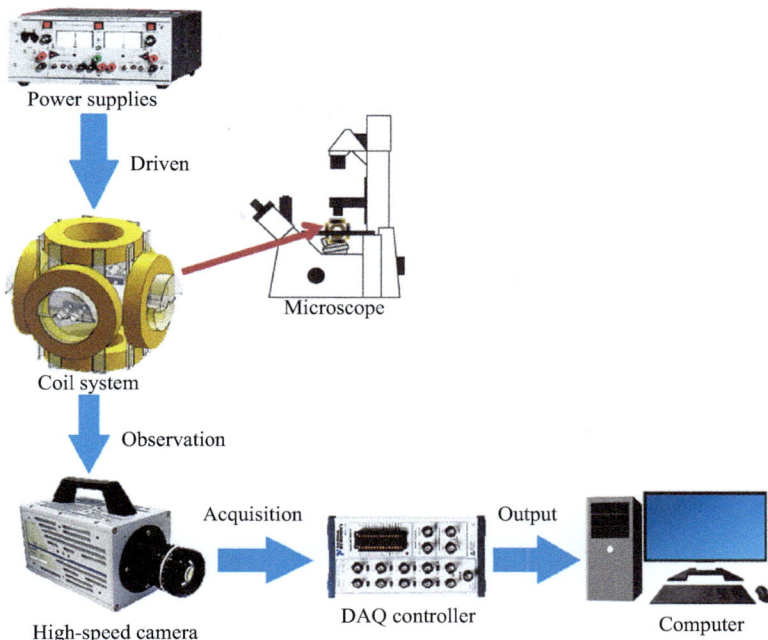

Figure 3.10   Control system for wireless magnetic control

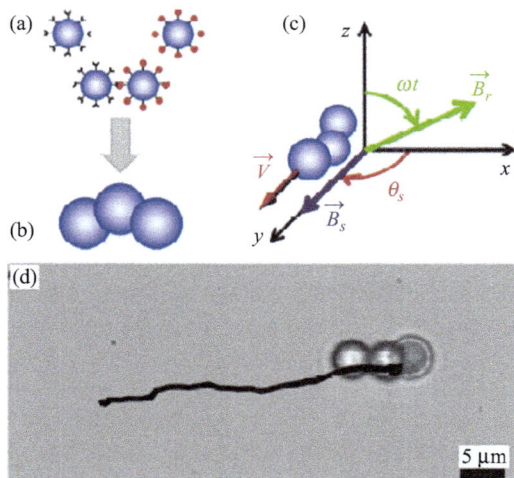

Figure 3.11   The achiral nanorobot: (a) avidin-coated and biotin-coated magnetic microbeads are mixed together to create; (b) three-bead achiral nanorobot; (c) a nanorobot and its swimming direction relative to the rotating magnetic field displayed in Cartesian coordinates; and (d) example of an achiral nanorobot swimming

intensity at incremental depths. Then, the intensity standard deviation of each bead of the magnetic particle was computed and matched onto the calibration curve to estimate the z positions of each bead. Achiral magnetic particles that underwent six cycles of rotations will experience less than 5% changes in the relative distances, signifying that deformation is minimal and that flexibility is a nonissue [4].

While the handedness has a physical implication to motion control, there is also uncertainty associated with it. Contrary to helical swimmers, an achiral nanorobot does not always swim in the reverse direction when the driving rotational magnetic field reverses the direction of its rotation. The motion uncertainty manifested in two modes of swimming motion: the primary motion and the secondary motion. Under the primary motion, a nanorobot exhibits steady forward swimming whose direction depends on the handedness, while under the secondary motion, a nanorobot can move forward or backward. Thus, it is necessary to introduce a control input to completely avoid the conditions for the secondary motion by maintaining the matched set for the primary motion [4].

The strength of the magnetic field generated from a pair of coils in the approximate Helmholtz (cube) configuration can be calculated using a modified version of the Biot–Savart law [36] which is given as

$$B_{coils} = \frac{\mu_0 nIR^2}{2(R^2 + x^2 - 2dx + d^2)^{\frac{3}{2}}} \pm \frac{\mu_0 nIR^2}{2(R^2 + x^2 + 2dx + d^2)^{\frac{3}{2}}} \tag{3.16}$$

where μ_0 is the permeability, n is the number of turns of wires per coil, I is the electrical current passing through the wires, R is the effective radius of the coil, d is the distance between the center of a coil pair, and x is the coil distance to a point. The changes to the original Biot–Savart law is to specify a different distance (d) between a pair of coils, whereas the original equation set the distance (d) as the radius of the coil. Each of the terms in that equation represents one coil. The effective magnetic field is the resultant of coils' fields.

The secondary motion could be eliminated by using a combination of a rotating magnetic field and a weak static magnetic field. The magnetic fields are generated by the 3D approximate Helmholtz coil system, from which static field is used to orient the nanorobot in the desired direction while the rotating field is used to actuate the nanorobot to create propulsion.

With this experiment platform, it is possible to verify the model and conclusion from the simulation platform. For instance, to verify the movement of nanorobot in Figure 3.7(b) in Section 3.3.4, two schemes might be applied. Figure 3.12(a) shows the channel used in the first scheme. In this channel, the nanorobots are designed to take different routes to the end of a fixed target position. The direction of the guiding magnetic field is kept pointing to the target from the nanorobots' current location. Thus, the angle is different for each node on different routes. By comparing the number of nanorobots arrived from different routes and the theoretical calculation results, we could verify whether the control scheme can effectively guide the movement of nanorobots in environments with unclear resolution. The second scheme will create a single branch structure similar to Figure 3.12(b) and (c). By constantly adjusting the strength of the magnetic field and the angle, we expect to test the effects of the magnetic field in the same branch environment.

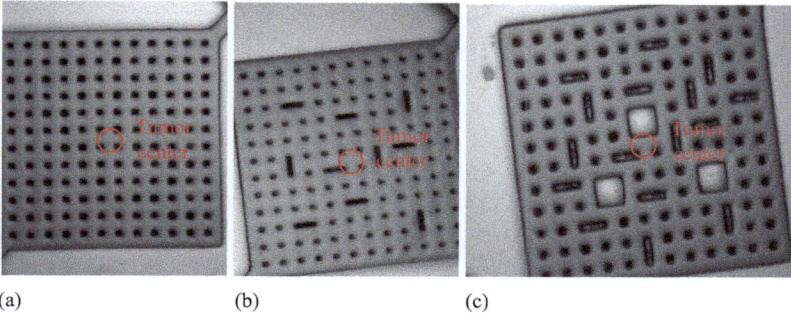

(a)   (b)   (c)

Figure 3.12   Simulated multi-layer vascular network with three different occupancy levels

References

[1]   S. W. Hwang, H. Tao, D. H. Kim, *et al.*, "A physically transient form of silicon electronics," *Science*, vol. 337, pp. 1640–1644, 2012.

[2]   J. L. Ramos, E. Díaz, D. Dowling, *et al.*, "The behavior of bacteria designed for biodegradation," *Nature Biotechnology*, vol. 12, pp. 1349–1356, 1994.

[3]   S. Martel, M. Mohammadi, O. Felfoul, *et al.*, "Flagellated magnetotactic bacterial as controlled MRI-trackable propulsion and steering systems for medical nanorobots operating in the human microvasculature," *International Journal of Robotics Research*, vol. 28, pp. 571–582, 2009.

[4]   U. K. Cheang, H. Kim, D. Milutinovi, *et al.*, "Feedback control of an achiral robotic microswimmer," *Journal of Bionic Engineering*, vol. 14, pp. 245–259, 2017.

[5]   S. Hiyama, Y. Moritani, T. Suda, *et al.*, *Molecular Communication*, Springer, London, 2005, pp. 205–228.

[6]   R. P. Feynman, "There's plenty of room at the bottom," *Journal of Microelectromechanical Systems*, vol. 1, no. 1, pp. 60–66, 1992.

[7]   Y. H. Bae and K. Park, "Targeted drug delivery to tumors: Myths, reality and possibility," *Journal of Controlled Release*, vol. 153, no. 3, pp. 153–198, 2011.

[8]   C. Liu, T. Xu, L. P. Xu, and X. Zhang, "Controllable swarming and assembly of micro/nanomachines," *Micromachines*, vol. 9, no. 1, pp. 9–10, 2017.

[9]   F. Guo, P. Li, J. B. French, *et al.*, "Controlling cell–cell interactions using surface acoustic waves," *Proceedings National Academy of Science*, vol. 112, no. 1, pp. 43–48, 2015.

[10] Y. Chahibi, M. Pierobon, and I. F. Akyildiz, "Pharmacokinetic modeling and biodistribution estimation through the molecular communication paradigm," *IEEE Transactions on Biomedical Engineering*, vol. 62, no. 10, pp. 2410–2420, 2015.

[11] D. A. T. Pizzo, "1906.1-2015—IEEE Recommended Practice for Nanoscale and Molecular Communication Framework," 2016.

[12]   S. Martel, O. Felfoul, J.-B. Mathieu, *et al.*, "MRI-based medical nanorobotic platform for the control of magnetic nanoparticles and flagellated bacteria for target interventions in human capillaries," *International Journal on Robotics Research*, vol. 28, pp. 1169–1182, 2009.

[13]   S. P. Sherlock and H. Dai, "Multifunctional FeCo-graphitic carbon nanocrystals for combined imaging, drug delivery and tumor-specific photothermal therapy in mice," *Nano Research*, vol. 4, pp. 1248–1260, 2011.

[14]   I. L. Medintz, H. Tetsuo Uyeda, E. R. Goldman, and H. Mattoussi, "Quantum dot bioconjugates for imaging, labeling and sensing," *Nature Materials*, vol. 4, pp. 435–446, 2005.

[15]   Y. Chen and P. Kosmas, "Detection and localization of tissue malignancy using contrast-enhanced microwave imaging: Exploring information theoretic criteria," *IEEE Transactions on Biomedical Engineering*, vol. 59, pp. 766–776, 2012.

[16]   Y. Chen, P. Kosmas, P. S. Anwar, and L. Huang, "A touch-communication framework for drug delivery based on a transient microbot system," *IEEE Transactions on Nanobioscience*, vol. 14, no. 4, pp. 397–408, 2015.

[17]   Y. Zhou, Y. Chen, R. D. Murch, R. Wang, and Q. Zhang, "Simulation framework for touchable communication on NS3Sim," *Nano Communication Networks*, vol. 16, pp. 26–36, 2018.

[18]   M. Zamir, "Fractal dimensions and multifractility in vascular branching," *Journal Theoretical Biology*, vol. 212, pp. 183–190, 2001.

[19]   E. Gabryś, M. Rybaczuk, and A. Kędzia, "Fractal models of circulatory system. Symmetrical and asymmetrical approach comparison," *Chaos, Solitons & Fractals*, vol. 24, pp. 707–715, 2005.

[20]   M. Rispoli, M. C. Savastano, and B. Lumbroso, "Capillary network anomalies in branch retinal vein occlusion on optical coherence tomography angiography," *Retina*, vol. 35, pp. 2332–2338, 2015.

[21]   I. L. Medintz, H. Tetsuo Uyeda, E. R. Goldman, and H. Mattoussi, "Quantum dot bioconjugates for imaging, labeling and sensing," *Nature Materials*, vol. 4, pp. 435–446, 2005.

[22]   Y. Gazit, D. A. Berk, M. Leunig, L. T. Baxter, and R. K. Jain, "Scale-invariant behavior and vascular network formation in normal and tumor tissue," *Phys. Rev. Lett.*, vol. 75, no. 12, pp. 2428–2431, 1995.

[23]   Y. Chen, S. Shi, X. Yao, *et al.*, "Touchable computing: Computing-inspired bio-detection," *IEEE Transactions on NanoBioscience*, vol. 16, no. 8, pp. 810–821, 2017.

[24]   C. Liu, T. Xu, L. P. Xu, and X. Zhang, "Controllable swarming and assembly of micro/nanomachines," *Micromachines*, vol. 9, no. 1, pp. 9–10, 2017.

[25]   F. Guo, P. Li, J. B. French, *et al.*, "Controlling cell–cell interactions using surface acoustic waves," *Proceedings of National Academy of Sciences*, vol. 112, no. 1, pp. 43–48, 2015.

[26]   Y. Zhou, Y. Chen, and R. Murch, "Latticed channel model of touchable communication over capillary microcirculation network," *GLOBECOM 2018—2018 IEEE Global Communications Conference*, December 2018.

[27] H. Lee, A. M. Purdon, V. Chu, and R. M. Westervelt, "Controlled assembly of magnetic nanoparticles from magnetotactic bacteria using microelectro-magnets arrays," *Nano Letters*, vol. 4, no. 5, pp. 995–998, 2004.

[28] E. Glenn, *Viscosity*, The Physics Hypertext Book, pp. 2–5, 1998. Available from: https://physics.info/viscosity/.

[29] E. N. Marieb and K. Hoehn, "The cardiovascular system: Blood vessels," *Human Anatomy & Physiology* (9th ed.), Boston, MA: Pearson Education, p. 712, 2013.

[30] Y.-c. Fung and B. W. Zweifach, "Microcirculation: Mechanics of blood flow in capillaries," *Annual Review of Fluid Mechanics*, vol. 3, pp. 189–210, 1971.

[31] S. P. Sherlock and H. Dai, "Multifunctional FeCo-graphitic carbon nano-crystals for combined imaging, drug delivery and tumor-specific photothermal therapy in mice," *Nano Research*, vol. 4, pp. 1248–1260, 2011.

[32] B. J. Kirby, *Micro- and Nanoscale Fluid Mechanics: Transport in Microfluidic Devices*, New York: Cambridge University Press, 2010.

[33] D. W. McRobbie, *MRI from Picture to Proton*, Cambridge University Press, Cambridge, 2007.

[34] K. E. Peyer, L. Zhang and B. J. Nelson, "Bio-inspired magnetic swimming microrobots for biomedical applications," *Nanoscale*, vol. 5, pp. 1259–1272, 2013.

[35] T. R. Kline, W. F. Paxton, T. E. Mallouk, *et al.*, "Catalytic nanomotors: Remote-controlled autonomous movement of striped metallic nanorods," *Angew Chemical International Edition*, vol. 44, pp. 744–746, 2005.

[36] D. J. Griffiths, *Introduction to Electrodynamics*, Prentice Hall, Upper Saddle River, NJ, 1999.

Part II

Current development in THz components and interfaces

Chapter 4

Terahertz antenna design for wearable applications

*Abdel Baset[1], Muhammad Ali Imran[1], Akram Alomainy[2]
and Qammer H. Abbasi[1]*

4.1 Introduction

Terahertz (THz) waves, also known as sub-millimetre waves, are observed between the optical and microwave frequency regions of the electromagnetic spectrum [1,2]. Devices operating in the optical and infrared spectra are generally characterized by electromagnetic beams that contain many modes, and the dimensions of such devices are typically much larger than the operating wavelength. On the contrary, THz frequency-based devices are comparable in size to the operating wavelength [3]. Such devices have the potential to fulfill the future wireless communications needs of ultra-high bandwidths and data rates. Moreover, the problems of channel congestion can be overcome through the use of THz-based communication systems. There has been an ever-increasing demand of the data-based wireless communication in the past few decades. Moreover, throughout the world, the urban population has been growing at an alarming rate. These two factors have created enormous challenges in implementing high-speed and reliable networks that can serve a large number of people at the same time [4]. A natural solution to overcome this challenge is to shift the network operating frequency to the THz region. However, with the available technologies, THz systems are affected by high atmospheric attenuation and, therefore, much lower propagation distances. Hence, there is a demand for novel techniques through which the performance of a THz-based communication system can be increased. To do so, micro- and nanoscale device fabrication techniques have actively been studied of late. Additionally, the search for highly conductive materials is an active research area to develop efficient THz devices. In conventional wireless communication systems, metallic antennas are fabricated separately from the microelectronics. It is not currently

[1]James Watt School of Engineering, University of Glasgow, Glasgow, UK
[2]School of Electronic Engineering and Computer Science, Queen Mary University of London, London, UK

possible to operate circuitry in the THz frequency region simply by down-scaling the traditional metallic antenna to a few micrometres [5]. By moving up in the frequency, the device physics changes drastically and the approach of using a metal-based antenna has several limitations, chief among them is the low mobility of electrons in the nanoscale metallic structures. This results in the antennas being highly lossy at the resonant frequencies, which would result in a high attenuation and, subsequently, poor efficiency of the overall system [6]. To address this lossy behavior, meta-material-based nano-structures have emerged as attractive solutions [7]. Graphene is an atomically thin, two-dimensional crystalline form of carbon in which the carbon atoms are arranged in a hexagonal lattice structure. It was discovered in 2004 by Novoselov and Geim, who peeled off a small amount of monolayer graphene with the help of adhesive tape, therefore, obtaining free-standing graphene in air [8]. For their ground-breaking discovery, they were awarded the Nobel Prize in Physics in 2010. Graphene is considered one of the thinnest materials yet, but also the strongest one measured in terms of mechanical properties [9,10]. Its theoretical specific surface area is as high as 2,630 m^2 g^{-1}, thermal conductivity as high as 5,300 W m^{-1} K^{-1}, which is higher than that of the carbon nanotube and diamond. In terms of electromagnetic properties, it is almost completely transparent, absorbing only 2.3% of light. On the other hand, graphene also has a very high electron mobility of 15,000 cm^2 V^{-1} S^{-1} at room temperature, which is higher than that of the carbon nanotube or crystalline silicon [11]. Nonetheless, the electrical resistivity of graphene is of the order of 1 × 10^{-6} Ω cm lower than copper and silver, which are considered the best conducting materials. All these attractive characteristics enable graphene to display extraordinary electromagnetic phenomena. When an electromagnetic wave, especially in the THz frequency range, is obliquely incident wave upon graphene, surface plasmon polaritons (SPPs) are excited. Due to the high conductivity of graphene, the SPPs are tightly confined to the graphene surface having a wavelength much smaller than its free-space equivalent. Moreover, the graphene SPPs exhibit moderate loss, and a vital property of tuning of the resonant frequency through external electrical and magnetic bias, and chemical doping. Most importantly, the resonant frequency of graphene lies in the THz and mid-infrared frequency ranges. Hence, this property, coupled with the lightweight and the flexible nature of graphene, opens up the possibility of creating miniaturized and flexible antenna devices operating in the THz frequency domain that is well suited for wearable applications.

The electrical conductivity of graphene in the THz frequency range can be described using a semi-classical electronic model in the absence of magnetic bias. It is typically expressed in terms of the well-known frequency-dependent Kubo formula [12]

$$\sigma(\omega) \approx -j\frac{e^2 k_B T}{\pi \hbar^2 (\omega - j\tau^{-1})} \times \left(\frac{\mu_c}{k_B T} + 2\ln(e^{-\mu_c/(k_B T)} + 1)\right) \tag{4.1}$$

where e is the electron charge, τ is the electron relaxation time in graphene, k_B is the Boltzmann constant, T is the temperature, $\hbar$ is the reduced Planck constant, and

μ_c is the chemical potential of graphene. The interband transitions [imaginary part in (4.1)] only occur when $\hbar\omega < 2\mu_c$, which is the case while operating in the THz frequency range for moderate μ_c. For graphene, one of the most significant characteristics is the ability to control or, in other words, tune the complex-valued conductivity. This can be done by exploiting the field effect with an application of an external electrostatic field bias. The opto-electronic response of graphene can be described in terms of μ_c and τ as [13]

$$\mu_c = \hbar v_f \sqrt{\pi N}, \tag{4.2a}$$

$$\tau = \frac{\mu_c \mu}{e v_f^2} \tag{4.2b}$$

where v_f is the Fermi velocity, μ is the carrier mobility of the electrons, and N is the carrier density. When μ_c and N are low, the complex-valued conductivity (4.1) has smaller real and imaginary parts. This leads to poor efficiency in terms of radiation as the SPPs have a lower propagation length along the graphene surface. Therefore, in order to observe a strong resonant behavior and improve efficiency, μ_c needs to be enhanced, which is usually done through chemical doping. Although the increase in both μ_c and μ leads to efficient radiation, the effect on the resonant frequency is different in both cases. Moreover, an increase in the radiated energy affects the width of the resonant frequency response introducing ringing tails around the centre frequency. Such broadening of the temporal response is coherent with the sharpening of the resonant behavior that is observed when the μ is increased. On the other hand, a stronger resonant behavior is observed when τ is increased.

On the other hand, perovskite, which was first discovered in the Ural Mountains and named after a Russian scientist, Lev Perovskite [14], is a mineral with a unique crystalline structure. A perovskite structure is any compound that has a similar structure to the perovskite mineral. True perovskite (the mineral) is made of calcium, titanium, and oxygen with the structure $CaTiO_3$. Perovskites have been widely inspected all around the world, given their attractive properties, especially in regard to their photovoltaic and plasmonic applications [15]. Perovskite materials display many outstanding characteristics in terms of superconductivity, ferroelectricity and high thermo-power [16]. Recently, there has been an increased interest in the lead halide-based perovskite-type crystals owing to their extraordinary optical properties. They are self-organized low-dimensional crystals. However, $CH_3NH_3PbI_3$-type perovskites have been recognized as promising materials, having a high carrier mobility and diffusion length, indicating that these materials can be to design antennas. Despite the attractive properties of perovskites, controlling the physical properties during the design and fabrication phase is still quite challenging, which is currently a limitation in the realization of the next-generation functional devices meant for high frequencies [17]. Moreover, various environmental factors, including moisture, ultraviolet light exposure, and thermal stress, play a key role in the instability of perovskite materials [18].

4.2   Graphene-based antenna for wearable applications

A full-wave 3D electromagnetic simulation tool, CST Microwave Studio 2018, was used to analyze different antenna designs based on graphene acting as the main radiating element. All simulations were done at room temperature (293 K). In the first design, a patch antenna, as shown in Figure 4.1, was analyzed, which consisted of graphene as the conducting material and a dielectric substrate. The effect of μ_c and τ on the radiation performance of the patch antenna was assessed by varying it in the range of 0.1–0.4 eV and 0.1–0.8 ps. The return loss of the patch antenna in the THz frequency range is shown in Figure 4.2. It is noted that as τ increased, the higher-order modes of the fundamental resonant frequency become prevalent.

The strongest resonance having the smallest value of (S_{11}), −45 dB, is obtained at a chemical potential of 0.4 eV and 0.5 ps relaxation time. The resonant frequency of the graphene also changes when the chemical potential is varied. The effect is shown in Figure 4.2. When μ_c has a value of 0.4 eV, the resonant frequency is observed at 4.636 THz, whereas at 0.2 eV, it is 4.546 THz. Furthermore, S_{11} started to increase from −34 to almost −37 dB. At 0.4 eV, S_{11} is still below −10 dB at 0.1 ps relaxation time, and get greater reflection coefficient, relaxation time must be increased, as a result, S_{11} increased to −45 dB when relaxation time reaches 0.5 ps, the resonant frequency shifted to 5.3 THz as an alternative of 4.7 THz at 0.3 eV. The antenna has three possible resonant frequencies, as shown in Table 4.1. The thickness of the substrate is evaluated to optimize the antenna performance in terms of the reflection coefficient and bandwidth. From Figure 4.3, the substrate thickness obviously affects the S_{11} value; however, the bandwidth becomes narrower with increasing substrate thickness (Table 4.2).

As shown by the results, a substrate thickness of 7 μm generates the best resonance with an S_{11} of −41 dB and a bandwidth of 192 GHz. The selection of the substrate thickness is, therefore, a critical design parameter, which strongly affects

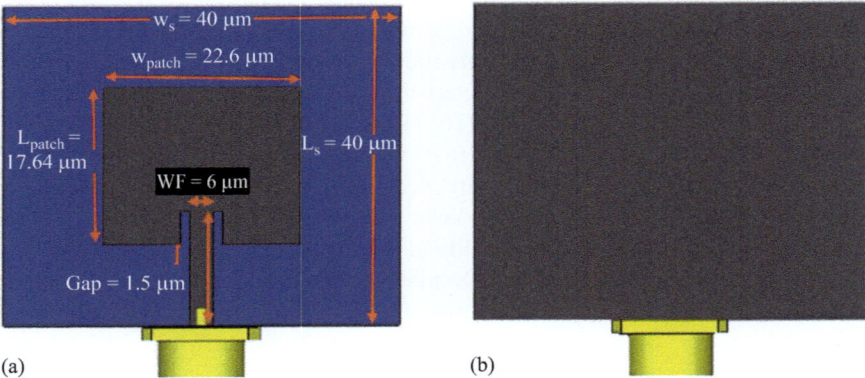

(a)   (b)

Figure 4.1   Patch antenna with a flexible substrate and graphene as the radiating element: (a) front view and (b) back view

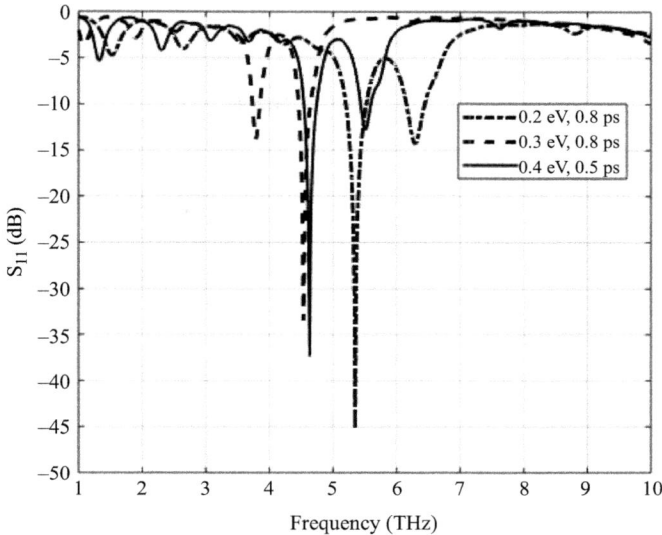

Figure 4.2 *The effect of chemical potential observed on the reflection coefficient (S_{11}) in the THz frequency range. The value of μ_c varied to 0.2 eV, 0.3 eV and 0.4 eV brown, whereas τ varied from 0.1 ps to 0.5 ps*

Table 4.1 *The effect of chemical potential and relaxation time on resonant frequency and bandwidth*

f (THz)	μ_c (eV)	τ (ps)	Bandwidth (GHz)
4.546	0.2	0.8	185
4.636	0.3	0.8	204
5.347	0.4	0.5	310

the antenna resonance. For wearable electronic applications, a substrate material having flexibility is an essential requirement. In order to compare the performance of different types of flexible substrate materials, μ_c of 0.2 eV and τ equal to 0.8 ps are selected. From Figure 4.4, the polyamide substrate (dielectric constant, $\varepsilon_r = 4.5$, and loss tangent, $\tan \alpha = 0.0027$) performs the best in terms of the S_{11} of -42 dB. The polyamide substrate is appropriate for applications demanding a high degree of dimensional stability in extreme environmental conditions. Moreover, polyamide offers high flexibility and low profile making it the perfect substrate material for printed fabrication techniques. On the other hand, the Rogers 3006 substrate yielded a return loss of -40.6 dB, and for polyethylene terephthalate (PET), it was -30 dB. For paper, the return loss was -30 dB.

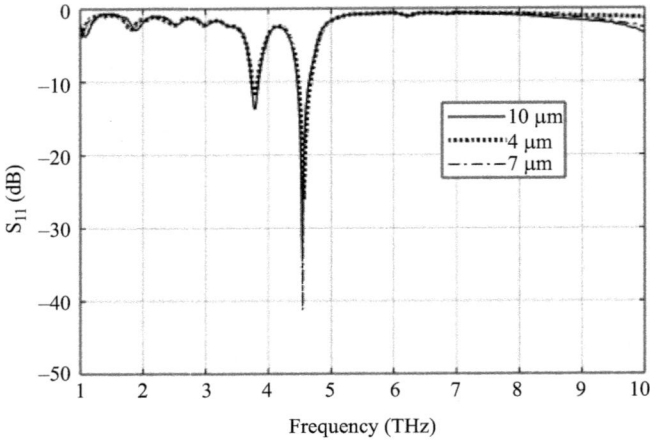

Figure 4.3   *Frequency sweep of the reflection coefficient for different values of substrate thickness*

Table 4.2   *The effect of substrate thickness on the reflection coefficient and bandwidth*

Substrate thickness (μm)	f (GHz)	Bandwidth (GHz)	S₁₁ (dB)
4	4.546	193.9	−26
7	4.546	192	−41
10	4.546	185	−33.3

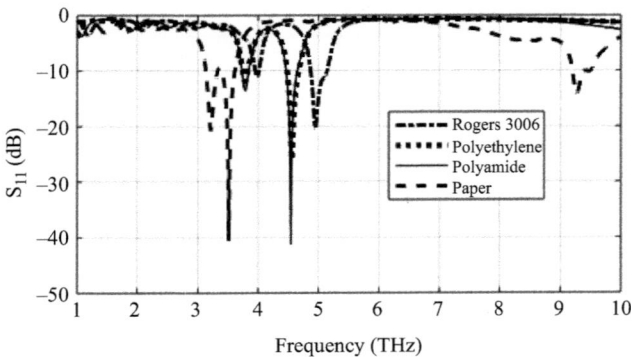

Figure 4.4   *Reflection coefficient (S_{11}) of different substrate material thickness 7 μm, Rogers 3006, polyethylene, polyamide, paper*

The tunability of graphene is shown in Figure 4.5, where the resonant frequency is changed with the help of chemical doping and applying an external voltage bias. Table 4.3 provides a comparison of the three different resonant frequencies obtained for a substrate of thickness 7 μm. All the frequencies have approximately the same S_{11} with the notable difference in bandwidth, which increases as the chemical potential increases. On the other hand, the transmission range gets lower when the electric potential is increased due to an increase in absorption. In Figure 4.6(a) and (b), the radiation patterns illustrate high main lobe magnitudes along with lower back lobe levels. The main lobe magnitude in the E-plane starts from 2.93 dB at a chemical potential of 0.2 eV, and with increased μ_c, the main lobe magnitude decreases to 1.51 dB at 0.3 eV and −1.41 dB at 0.4 eV. The proposed design suggests that the graphene-based patch antenna resonates at different frequencies in the THz band, 4.546, 4.636, and 5.347 THz when the chemical potential and relaxation time are varied. Furthermore, changing the chemical potential leads to an increase in bandwidth from 199 GHz at 0.2 eV to 314 GHz at 0.4 eV. On the other hand, chemical potential affects the radiation pattern by increasing the sidelobe and reducing the directivity of the proposed

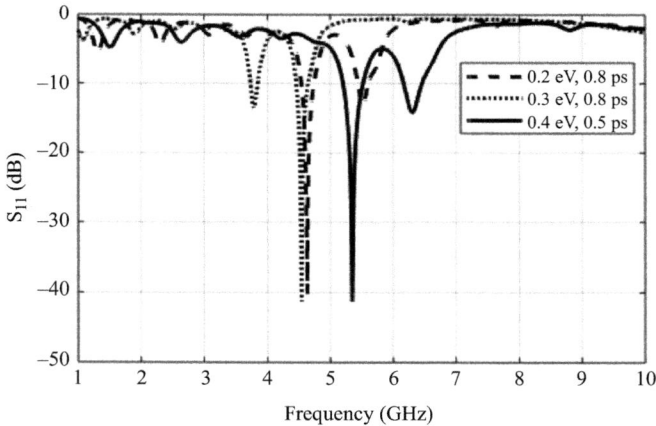

Figure 4.5   Antenna resonant frequencies at 7 μm thickness (0.2 eV and 0.8 ps, 0.3 eV and 0.8 ps, 0.4 eV and 0.5 ps)

Table 4.3   A comparison between different resonant frequencies at 7 μm substrate thickness

f (THz)	μ_c (eV)	τ (ps)	Reflection coefficient (S_{11})	Bandwidth (GHz)
4.546	0.2	0.8	−41.258	199
4.636	0.3	0.8	−40.187	279.9
5.347	0.4	0.5	−41.283	314

Farfield directivity abs (Phi = 90)

$(f = 4.546)$
$(f = 4.636)$
$(f = 5.347)$

Phi = 90 30   0   30 Phi = 270
60   60
90   90
120   120
150   180   150

Theta/degree vs. dBi

(a)

Farfield directivity abs (Phi = 0)

$(f = 4.546)$
$(f = 4.636)$
$(f = 5.347)$

Phi = 0   30   0   30 Phi = 180
60   60
90   90
120   120
150   180   150

Theta/degree vs. dBi

(b)

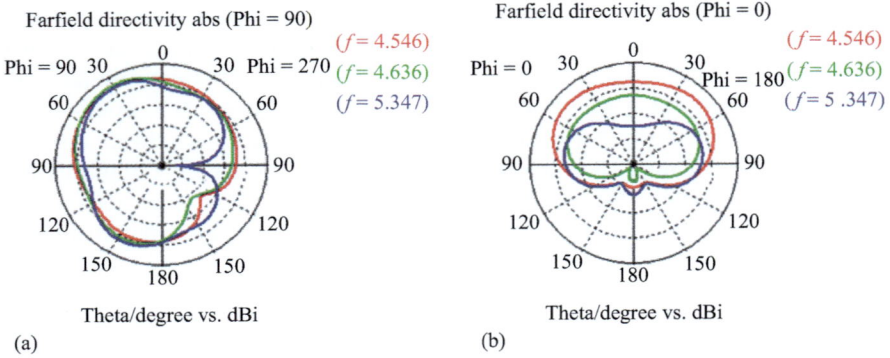

Figure 4.6   Simulations of the normalized field patterns in the E-plane (a) and H-plane (b) at different frequencies

antenna. Analysis has also been performed to evaluate the antenna bandwidth and reflection coefficient (S_{11}) at resonant frequencies. A comparison between different flexible substrates allows us to evaluate the effect of the substrate material on the antenna performance. A graphene-based antenna with a polyamide substrate shows the maximum S_{11} of −42 dB.

4.3   The effect of human body on the antenna radiation characteristics

The demand for wearable devices is expected to increase to 187.2 million wearable units annually by 2022 [19]. Wearable antenna requirements for all modern applications include lightweight, low-cost, and a flexible profile. For wireless body area network scenarios, the antenna design becomes more complicated than free-space environments, due to the absorption of the human skin. Human skin is a complex heterogeneous and anisotropic medium, where minuscule organs such as blood vessel and pigment content are spatially distributed in depth [20]. With the complexity of human skin, it is challenging to accurately describe the material, mainly due to the shapes and functions, and most importantly because of the absence of the dielectric constant measurements at high frequency [21]. Therefore, most of the latest research simulate the human skin as three layers; epidermis, dermis, and hypodermis which represent the most essential parts of the human skin [22]. Wearable antennas, therefore, should be carefully designed to minimize skin absorption. Figure 4.7(a) shows the geometry and parameters of the proposed co-planar waveguide-fed antenna. The antenna design is simulated at room temperature (293 K). The antenna design has the dimensions of 260 μm × 195 μm × 0.35 nm, with two layers of graphene, and consists of a graphene-based rectangular patch with a feedline 120 μm × 100 μm × 0.35 nm made from graphene. Rogers 3006, having a

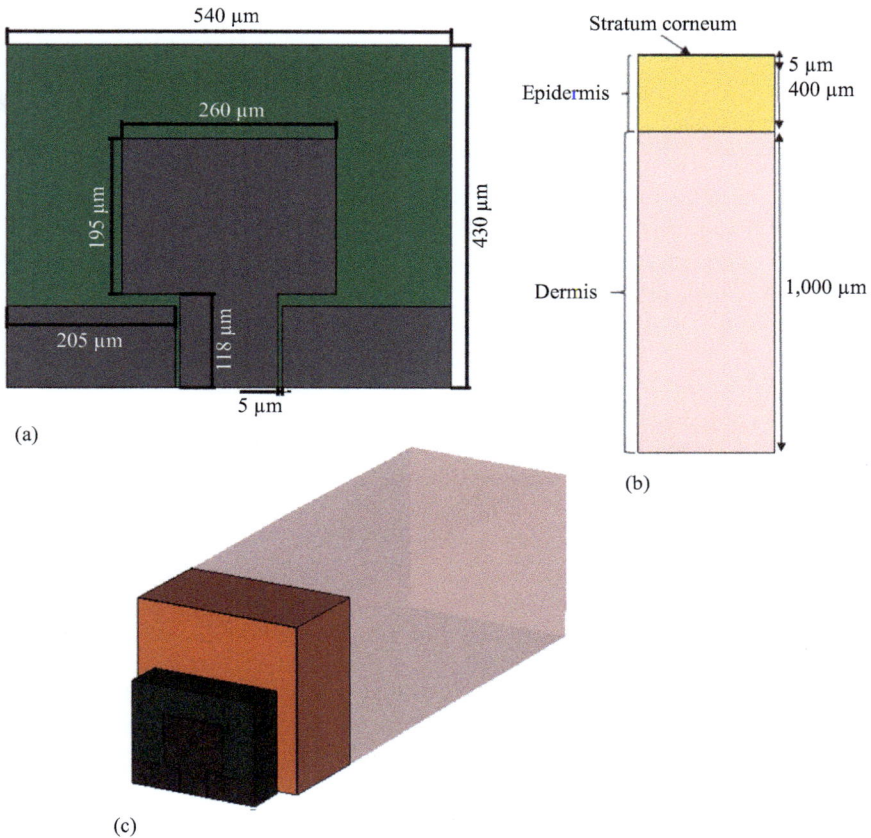

*Figure 4.7   Geometry of the proposed patch antenna. (a) Front view,
(b) human skin model, and (c) side view of the antenna on the
human body*

dielectric constant of $\varepsilon_r = 6.5$, and loss tangent, $\tan \alpha = 0.002$, is used as a substrate with a thickness of 175 μm. Three layers of the human skin model is shown in Figure 4.7(b), and the thickness of which differs between various human skins. For the epidermis, the typical thickness ranges from 0.05 to 1.5 mm, whereas the dermis is typically 1.5×10^{-4} mm. The hypodermis, on the other hand, has no typical value [23]. The epidermis is subdivided into two further layers, stratum corneum with only dead squamous cells and the living epidermis layer, where most of the skin pigmentation stay [13]. The stratum corneum is a thin accumulation on the outer skin surface. The dermis that supports the epidermis is thicker and mainly composed of collagen fibres and intertwined elastic fibres enmeshed in a gel-like matrix. The subcutaneous fat layer is composed of the packed cells with considerable fat, where the boundary is not well defined, and, thus, the thickness of this layer differs widely for various parts of the human body. The real (ε^-) and imaginary $(\varepsilon^=)$ parts of the permittivity of the human skin tissues can be obtained using [24,25],

$$\varepsilon^- = n^2 - k^2 \tag{4.3a}$$

$$\varepsilon^= = 2nk \tag{4.3b}$$

where n is the refractive index and k is the extinction coefficient. The refractive indices for blood, skin, and fat are 1.97, 1.73, and 1.58, respectively [25]. The extinction coefficient is calculated using the measured absorption coefficient data available in [24,26] through

$$k = \frac{\alpha\lambda_0}{4\pi} \tag{4.4}$$

where λ_0 is the free-space wavelength and α is the absorption coefficient.

Figure 4.8 demonstrates the scattering response of the graphene antenna in free space. It is observed that the resonant frequency can be downshifted with an increase in the patch length, in this case, from 195 to 215 μm.

The S_{11} of the graphene patch antenna is −46 dB at 0.647 THz obtained with a length of 195 μm. A wide bandwidth of 29.2 GHz is also obtained, which is an added advantage for wearable applications. Next, a comparison of antenna performance parameters is presented in a free-space environment and the on-body condition. From Figure 4.9, the reflection coefficient of the patch antenna in the on-body state is shifted slightly toward the right side of the 0.6482 THz resonant frequency. The S_{11} of the graphene patch antenna for the on-body case is −25 dB. The shift in frequency is due to the high dielectric constant property of three layers

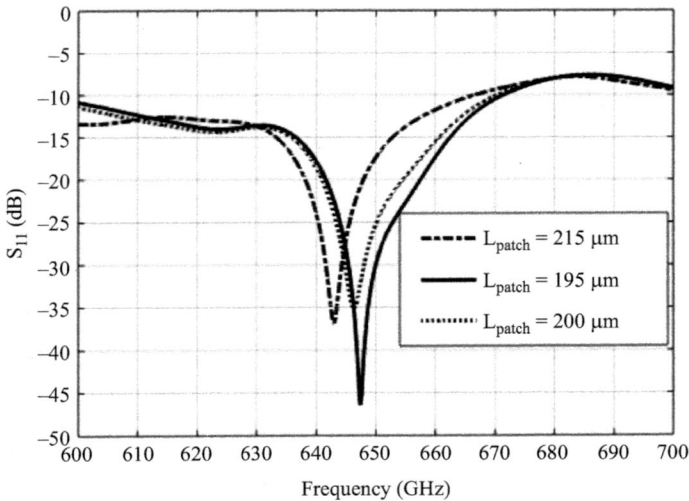

Figure 4.8 S_{11} of the antenna with different patch lengths

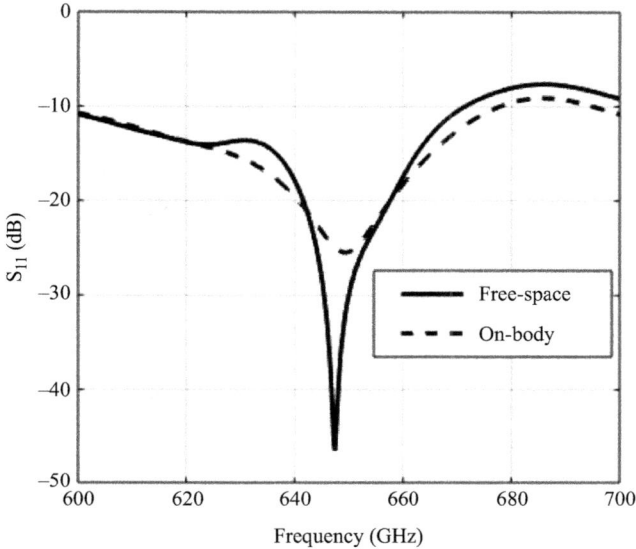

Figure 4.9 S_{11} for the on-body and free-space conditions

of the human body in proximity to the antenna. Because of these properties of the human body, most of the radiated waves propagate through the body and dissipate in the form of heat resulting in a wider −10 dB bandwidth. Due to the body absorption, the antenna gain decreases from −7.8 to −7.2 dB (Figure 4.10). The antenna gain is a figure of merit of how well the antenna converts the power supplied into radiated waves in a specific direction. For the case of the on-body condition, a lower value of the gain is obtained which is due to a decrease in the radiation efficiency, down from 96% to almost 50% (Figure 4.11). Therefore, the total radiated efficiency of the antenna on a flat body phantom decreases by nearly a factor of two. This is due to the higher conductivity of the outermost skin layer. The total antenna efficiency in the presence of the human body also decreases due to absorption in the lossy human body tissues.

The H- and E-plane radiation patterns of the graphene patch antenna in the on-body and free-space conditions are shown in Figure 4.12. The main lobe with −6.87 dB is obtained at a resonance of 0.647 THz, while an increase in the magnitude of the main lobe is observed at the on-body state with 9.7 dB. Similarly, the sidelobe in the on-body case is −8 dB and on the free-space is −2 dB due to higher reflections from the body surface. A wearable graphene antenna is presented and tested on three layers of human skin to serve the wearable applications in modern THz systems to achieve high performance. The results show that the designed antenna presented at 0.647 THz has a realized gain of 7.8 and 7.2 dB in the free-space and on-body cases, respectively. The radiation efficiency dropped from 96% to 50% when placed on the body.

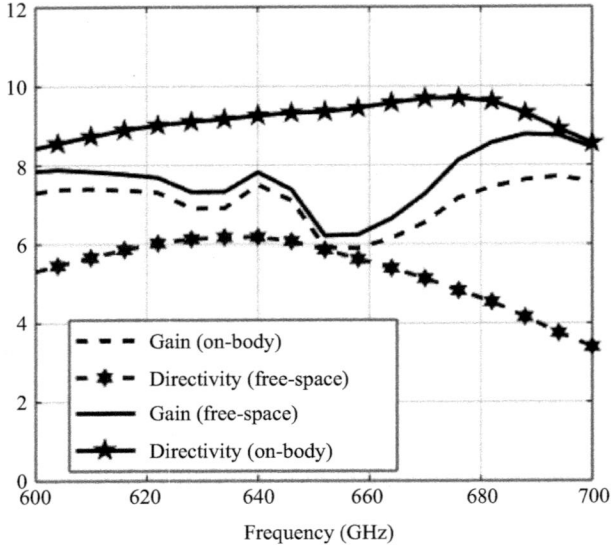

Figure 4.10   Simulation gain and directivity in free-space and on-body conditions

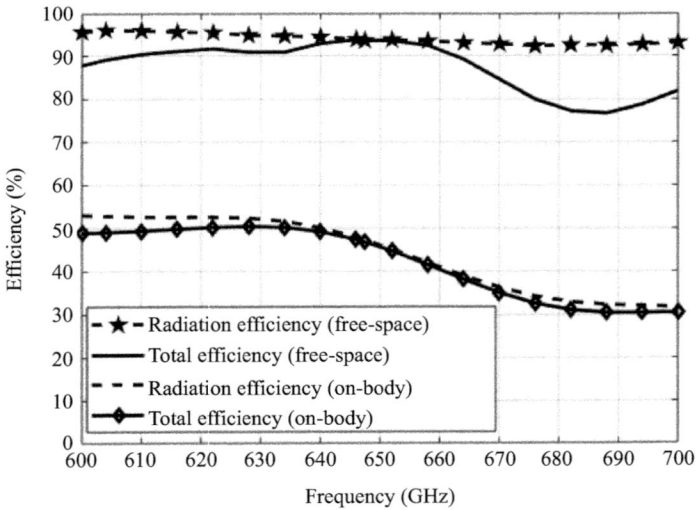

Figure 4.11   Radiation and total efficiency in free-space and on-body conditions

Farfield directivity abs (Phi = 90)

(a)

Farfield directivity abs (Phi = 0)

(b)

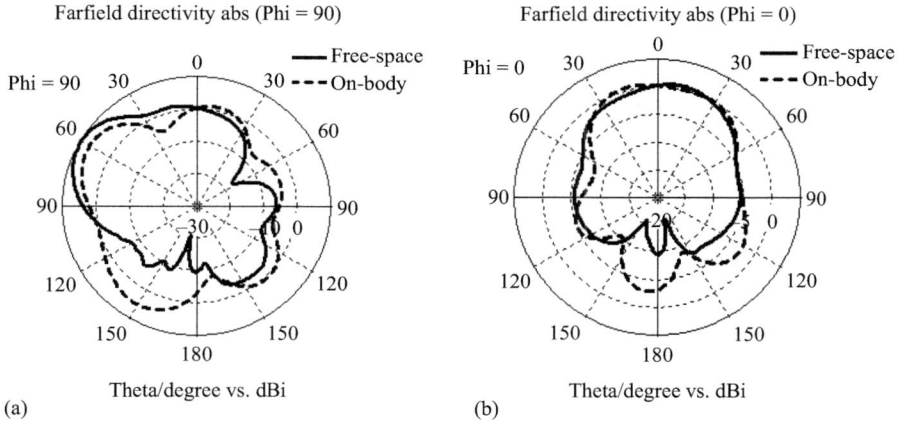

Figure 4.12   The (a) H-plane and (b) E-plane radiation patterns of the graphene patch antenna in the on-body and free-space conditions

4.4   Perovskite-based antennas

With the increasing demands of high speed, reliable and secure wireless communication and safety of the user in terms of radiation exposure, research trends have shown a migration toward the high-frequency THz spectrum [27,28]. Recent research efforts have been focused on overcoming the propagation and fabrication issues at THz and in developing THz sources, antennas, systems, and applications [29]. Therefore, THz devices can now be found in applications such as bio-sensing and imaging, fast and secure wireless systems and radar communication [30,31]. Due to high absorption levels of THz waves by the water molecules and limited penetration ability into the human body, the probability to produce hazardous ionization of biological tissues is low, which is an advantage of current medical imaging techniques such as X-rays. An efficient antenna is critical in developing a high-performance THz system. Therefore, several design aspects need to be considered to develop an antenna which is able to fulfil the demands of THz applications [32]. Rapid progress has been made in the utilization of low-dimensional materials moving on from the traditional copper-based radiating patch, which generates high loss and poor efficiency due to decrease in skin depth and conductivity in the THz frequency band. Carbon-based materials, i.e. graphene, and carbon nanotubes have shown promise in this regard [33]. Perovskites, due to the features such as superconductivity, ferroelectricity and low cost, have gained focus in various novel applications [34]. The proposed antenna design is numerically analyzed using CST Microwave Studio at room temperature (293 K). The perovskite material is simulated using the measured permittivity values as described in [18]. Figure 4.13 shows the antenna design with the dimensions 40 μm × 40 μm × 10 μm,

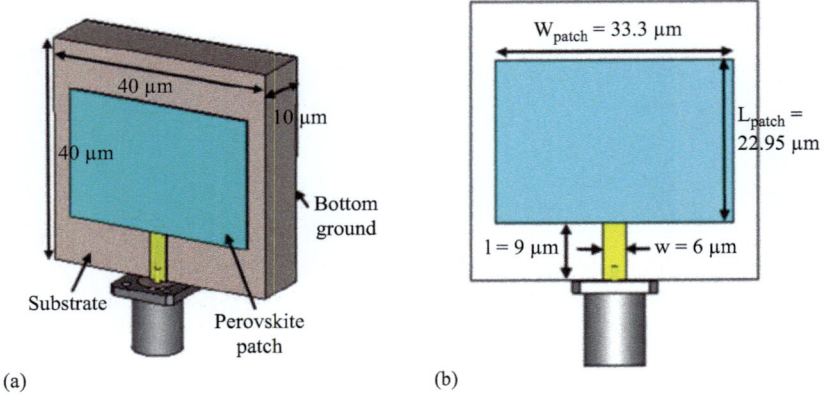

(a)   (b)

Figure 4.13   *Proposed perovskite-based THz antenna. (a) Simulated antenna design and (b) perovskite patch and copper feed line in the antenna prototype*

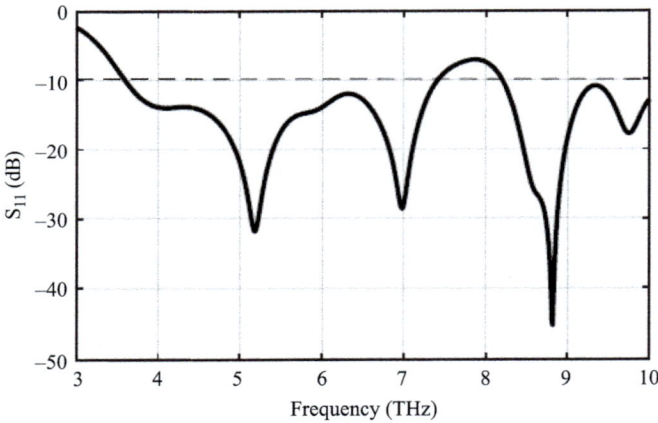

Figure 4.14   *Simulated S_{11} profile of the designed perovskite-based THz antenna*

consisting of a perovskite-based rectangular patch, while the ground plane and feedline are made of copper. A thin flexible film of polyamide is used as a substrate.

A frequency sweep of the reflection coefficient of the perovskite-based antenna is shown in Figure 4.14, which demonstrates two significant bands while taking −10 dB as a reference. The first band lies in the range of 3.6–7.4 THz and constitutes two dominant resonances at 5.18 and 6.98 THz. The second band covers 8.25–10 THz with a notable resonant dip at 8.8 THz. A parametric analysis is carried out on the radiating length of the antenna, and the results of which are shown in Figure 4.15. It is observed that with the increase in length from 20.95 to 29.7 μm, the resonant frequency is shifted down. The gain and efficiency versus

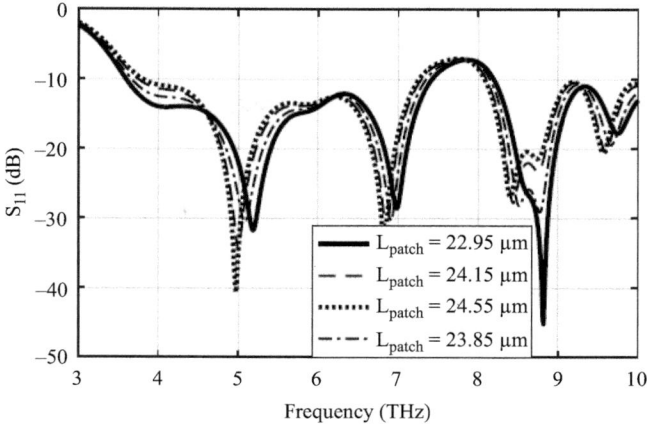

Figure 4.15 Parametric analysis of the radiating length of the designed perovskite-based THz antenna

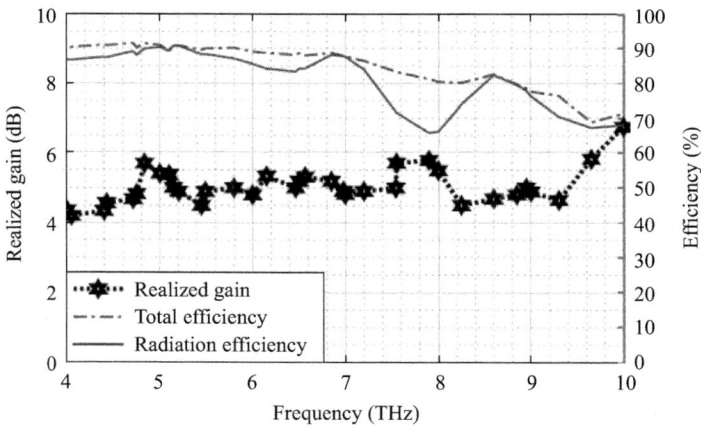

Figure 4.16 Realized gain and efficiency plots of the proposed perovskite-based THz antenna

frequency plots are shown in Figure 4.16, which indicate that the realized gain varies from −4.8 dB in the overall operating range. The gain magnitude is above 4 dB throughout the frequency sweep. Simulated results as shown in Figure 4.16 also present excellent efficiency of 88% at 5.185 THz, which is comparable to the reported graphene-based THz antenna in [23]. In addition, the radiation efficiency of the designed antenna can further be improved by increasing the conductivity of the perovskite film with the application of an external voltage. The E- and H-plane radiation patterns of the designed antenna at the three major resonant frequencies are illustrated in Figure 4.17. The main lobe with 4.53 dB is obtained at the

Farfield realized gain abs (Phi = 90)   Farfield realized gain abs (Phi = 0)

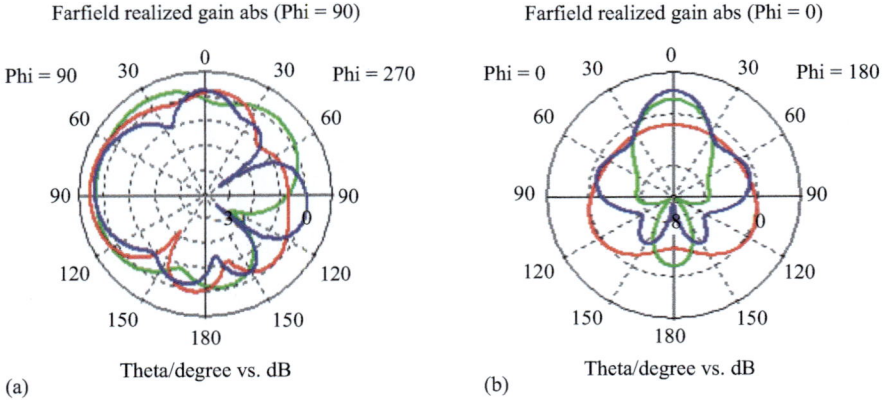

(a)   Theta/degree vs. dB   (b)   Theta/degree vs. dB

Figure 4.17   Proposed perovskite-based THz antenna. (a) E-plane and (b) H-plane

resonance of 5.185 THz, while an increase in the magnitude of the main lobe gain is observed at 6.985 and 8.821 THz resonances. Although the radiation is not perfect broadside in the complete operating range, this effect is obvious when the antenna is designed to cover a wide bandwidth. The high bandwidth, reasonable gain magnitudes, and high efficiency depict that the perovskite can be deployed as a radiating element in the THz patch antennas. The THz antennas are anticipated to ensure a high-resolution imagining for the biomedical applications due to their short wavelength and high directivity. Novel materials have been developed for potential deployment in modern THz systems to achieve high performance. A newly developed perovskite material has been used for the antenna design, which can replace the conventional metallic patches. The results show that the designed antenna covers two THz bands, 3.6–7.4 THz and 8.25–10 THz. The realized gain of the antenna is above 4 dB and the radiation efficiency is above 70% in overall operating bandwidth. The antenna is regarded as a potential candidate for the future short-range THz communications.

4.5   Conclusion

In this chapter, an overview of wearable antennas operating in the terahertz frequency range made from two-dimensional materials such as graphene is presented. The antenna designs are analyzed in realistic environments in the proximity of human skin. Characteristics such as highly miniaturized and flexible substrate materials of the antennas coupled with excellent antenna performance make these wearable antennas a strong candidate in applications of short-range wireless communication near the human body. The resonant properties of the two-dimensional materials are investigated using their electronic properties. Wireless communication in the terahertz frequency, high-resolution imaging for bio-sensing and disease management, and spectroscopy are anticipated to be some of the early beneficiaries of wearable and flexible antennas. Further investigations in this area of research provide interesting opportunities not only for antenna engineers but also for material scientists and physicists.

References

[1]   Duixian Liu, Ulrich Pfeiffer, Janusz Grzyb, and Brian Gaucher. *Advanced millimeter-wave technologies: antennas, packaging and circuits*. Chichester: John Wiley & Sons, 2009.

[2]   D Grischkowsky, Søren Keiding, Martin Van Exter, and Ch Fattinger. Far-infrared time-domain spectroscopy with terahertz beams of dielectrics and semiconductors. *JOSA B*, 7(10):2006–2015, 1990.

[3]   Jinhui Shi, Zhongjun Li, David K Sang, *et al*. Thz photonics in two dimensional materials and metamaterials: properties, devices and prospects. *Journal of Materials Chemistry C*, 6(6):1291–1306, 2018.

[4]   Ho-Jin Song and Tadao Nagatsuma. Present and future of terahertz communications. *IEEE Transactions on Terahertz Science and Technology*, 1(1):256–263, 2011.

[5]   Thomas Kleine-Ostmann and Tadao Nagatsuma. A review on terahertz communications research. *Journal of Infrared, Millimeter, and Terahertz Waves*, 32(2):143–171, 2011.

[6]   Javad Pourahmadazar and Tayeb A Denidni. Millimeter-wave planar antenna on flexible polyethylene terephthalate substrate with water base silver nanoparticles conductive ink. *Microwave and Optical Technology Letters*, 60(4):887–891, 2018.

[7]   Hou-Tong Chen, Willie J Padilla, Joshua MO Zide, Arthur C Gossard, Antoinette J Taylor, and Richard D Averitt. Active terahertz metamaterial devices. *Nature*, 444(7119):597, 2006.

[8]   Kostya S Novoselov, Andre K Geim, Sergei V Morozov, *et al*. Electric field effect in atomically thin carbon films. *Science*, 306(5696):666–669, 2004.

[9]   Jessica Campos-Delgado, José Manuel Romo-Herrera, Xiaoting Jia, *et al*. Bulk production of a new form of sp2 carbon: crystalline graphene nanoribbons. *Nano Letters*, 8(9):2773–2778, 2008.

[10]   Gwan-Hyoung Lee, Ryan C Cooper, Sung Joo An, *et al*. High-strength chemical-vapor–deposited graphene and grain boundaries. *Science*, 340(6136): 1073–1076, 2013.

[11]   Long Ju, Baisong Geng, Jason Horng, *et al*. Graphene plasmonics for tunable terahertz metamaterials. *Nature Nanotechnology*, 6(10):630, 2011.

[12]   Michele Tamagnone, Juan Sebastian Gomez-Diaz, Juan Ramon Mosig, and J Perruisseau-Carrier. Analysis and design of terahertz antennas based on plasmonic resonant graphene sheets. *Journal of Applied Physics*, 112(11): 114915, 2012.

[13]   Sergi Abadal, Ignacio Llatser, Albert Mestres, Heekwan Lee, Eduard Alarcón, and Albert Cabellos-Aparicio. Time-domain analysis of graphene-based

miniaturized antennas for ultra-short-range impulse radio communications. *IEEE Transactions on Communications*, 63(4):1470–1482, 2015.

[14]   Nam-Gyu Park, Tsutomu Miyasaka, and Michael Grätzel. *Organic-inorganic halide perovskite photovoltaics*. Cham, Switzerland: Springer, 2016.

[15]   Tze-Bin Song, Qi Chen, Huanping Zhou, *et al.* Perovskite solar cells: film formation and properties. *Journal of Materials Chemistry A*, 3(17):9032–9050, 2015.

[16]   Ani Kulkarni, Fabio T Ciacchi, Sarb Giddey, *et al.* Mixed ionic electronic conducting perovskite anode for direct carbon fuel cells. *International Journal of Hydrogen Energy*, 37(24):19092–19102, 2012.

[17]   Zubair Ahmad, Mansoor Ani Najeeb, RA Shakoor, *et al.* Instability in ch 3 nh 3 pbi 3 perovskite solar cells due to elemental migration and chemical composition changes. *Scientific Reports*, 7(1):15406, 2017.

[18]   Martin A Green, Anita Ho-Baillie, and Henry J Snaith. The emergence of perovskite solar cells. *Nature Photonics*, 8(7):506, 2014.

[19]   Guillermo Jesús Larios Hernández. Wearable technology: Shaping market opportunities through innovation, learning, and networking. In: Reyes-Mercado P., Larios Hernández G. (eds) *Reverse Entrepreneurship in Latin America*, pp. 29–44. Cham: Springer, 2019.

[20]   Cormac Flynn, Andrew Taberner, and Poul Nielsen. Modeling the mechanical response of in vivo human skin under a rich set of deformations. *Annals of Biomedical Engineering*, 39(7):1935–1946, 2011.

[21]   Stanislav I Alekseev and Marvin C Ziskin. Human skin permittivity determined by millimeter wave reflection measurements. *Bioelectromagnetics: Journal of the Bioelectromagnetics Society, The Society for Physical Regulation in Biology and Medicine, The European Bioelectromagnetics Association*, 28(5):331–339, 2007.

[22]   Samuel E Lynch, Robert B Colvin, and Harry N Antoniades. Growth factors in wound healing. single and synergistic effects on partial thickness porcine skin wounds. *The Journal of Clinical Investigation*, 84(2):640–646, 1989.

[23]   Mojtaba Dashti and J David Carey. Graphene microstrip patch ultrawide band antennas for THz communications. *Advanced Functional Materials*, 28(11):1705925, 2018.

[24]   Elizabeth Berry, Anthony J Fitzgerald, Nickolay N Zinov'ev, *et al.* Optical properties of tissue measured using terahertz-pulsed imaging. In *Medical Imaging 2003: Physics of Medical Imaging*, vol. 5030, pp. 459–471. International Society for Optics and Photonics, 2003.

[25]   Ke Yang, Alice Pellegrini, Max O Munoz, Alessio Brizzi, Akram Alomainy, and Yang Hao. Numerical analysis and characterization of THz propagation channel for body-centric nano-communications. *IEEE Transactions on Terahertz Science and Technology*, 5(3):419–426, 2015.

[26]   Anthony J Fitzgerald, Elizabeth Berry, Nickolay N Zinov'ev, *et al.* Catalogue of human tissue optical properties at terahertz frequencies. *Journal of Biological Physics*, 29(2–3):123–128, 2003.

[27]   William Ghann and Jamal Uddin. Terahertz (THz) spectroscopy: A cutting-edge technology. *Terahertz Spectroscopy A Cutting Edge Technology*, 2017. Available from: https://www.intechopen.com/books/terahertz-spectroscopy-a-cutting-edge-technology/terahertz-thz-spectroscopy-a-cutting-edge-technology.

[28]   Muhammad Saqib Rabbani and Hooshang Ghafouri-Shiraz. Liquid crystalline polymer substrate-based THz microstrip antenna arrays for medical applications. *IEEE Antennas and Wireless Propagation Letters*, 16:1533–1536, 2017.

[29]   Peter H Siegel. Terahertz technology. *IEEE Transactions on Microwave Theory and Techniques*, 50(3):910–928, 2002.

[30]   Leonardo Ranzani, Daniel Kuester, Kenneth J Vanhille, Anatoliy Boryssenko, Erich Grossman, and Zoya Popović. G-band micro-fabricated frequency-steered arrays with 2°/GHz beam steering. *IEEE Transactions on Terahertz Science and Technology*, 3(5):566–573, 2013.

[31]   Moustafa Khatib and Matteo Perenzoni. Response optimization of antenna-coupled FET detectors for 0.85-to-1-THz imaging. *IEEE Microwave and Wireless Components Letters*, 28(99):1–3, 2018.

[32]   Bolin Chen and Huang Fengyi. Research on a mems-technology-based corrugated horn antenna designed in tera-hertz regime. In *2016 IEEE International Conference on Microwave and Millimeter Wave Technology (ICMMT)*, vol. 1, pp. 213–216. IEEE, 2016.

[33]   Seyed Arash Naghdehforushha and Gholamreza Moradi. Design of plasmonic rectangular ribbon antenna based on graphene for terahertz band communication. *IET Microwaves, Antennas & Propagation*, 12(5):804–807, 2017.

[34]   Cheng-Liang Huang and Yu-Wei Tseng. A low-loss dielectric using catio 3-modified mg 1.8 ti 1.1 o 4 ceramics for applications in dielectric resonator antenna. *IEEE Transactions on Dielectrics and Electrical Insulation*, 21(5):2293–2300, 2014.

Chapter 5

Terahertz (THz) application in food contamination detection

Aifeng Ren[1,2], Adnan Zahid[2], Xiaodong Yang[1], Akram Alomainy[3], Muhammad Ali Imran[2] and Qammer H. Abbasi[2]

The electromagnetic (EM) radiation in the range of 0.3–3 THz has unique properties that make it particularly attractive for various applications including biomedical imaging, packaged goods inspection, and food and water contamination detection. However, generating radiation at a terahertz (THz) level presents many practical challenges. In recent years, several techniques for generating both continuous-wave and pulsed terahertz radiation have been developed [1]. In turn, these are spawning the early development of terahertz applications, particularly in microbial pollution in food inspection. New trends, discoveries, and applications have been observed in diverse fields, especially in the field of photonics and nanotechnology. The first demonstration of THz wave time-domain spectroscopy exploiting femtosecond laser sources to produce and identify freely transmitting THz pulses was developed by Auston at Bell Labs and by Grischkowsky at IBM in the 1980s [2]. The development of optimized techniques for THz pulse generations, particularly the optimized non-collinear beam geometry [3], as well as the production of THz pulses using the energy of few micro-joules or even higher values which has the potential to access the second- or third-order non-linearities, is the standard followed in almost every laboratory setting [4,5]. The highest-energy ultra-short THz pulses were generated by employing electro-optical rectification of femtosecond pulses in $LiNbO_3$ leveraging the tilted-pulse-front pumping technique, which is intrinsically expansible to boost the available THz pulse energy and corresponding field strength [6]. The THz radiation has many advantages over other imaging modalities in many aspects, including nonionizing [7], classifying species

[1]School of Electronic Engineering, Xidian University, Xi'an, Shaanxi, China
[2]Electronic and Nanoscale Engineering, University of Glasgow, Glasgow, UK
[3]School of Electronic Engineering and Computer Science, Queen Mary University of London, London, UK

of tissue [8], and scattering and water absorption for porosimetry [9]. Consequently, the THz radiation has shown widespread potential applications in diverse areas to address the real-world problems including genomics [10], medical diagnostics [11], pharmacology [12], healthcare applications [13,14], body-area networks [15,16], wireless communication [17,18], environmental monitoring [19], agriculture [20], food analysis [21,22], and defense and security [23,24].

The recent research on THz radiation generation, detection, and the spectroscopic and imaging tools, such as terahertz time-domain spectroscopy (THz-TDS) [25–27] which has the potential to be operated in the time and spatial domains to process the data obtained from the transmitted or reflected THz beam, opens up new ventures by applying the particular technology for food contamination detection [28]. The main idea of this application is that food and liquid have unique physical features and present a unique spectral imprint when exposed to the THz frequency domain [29–36]. The food quality and safety monitoring [31] comprising microbiological contamination detection, including toxic metals [36], pesticides [32], veterinary drug residues, organic pollutants, radionuclides, and mycotoxins, is a pressing issue for public health and well-being. Several studies have focused on different types of food additives, such as flour and talc mixtures [33] and melamine in milk powders [34], which indicate that the absorption spectra of the pure and mixed products were by and large distinct. The strong absorbing nature of THz radiation by water and the vibrational modes between molecules of hydrogen bonds positioned in the THz region present limitations on the application of THz technology considering the detection of concealed cases [37,38]. However, the absorbing nature does not pose a challenge to the applications of THz imaging in identifying humidity in different substances. Several methods that can resolve limited penetration depth issue have been provided, for example, the paraffin-embedding technique, the freezing technique, and terahertz penetration-enhancing agents (THz-PEAs) [11]. Some existing sensing techniques can be used for contamination detection at the THz level [39]. The existing studies on the THz sources and processing methods, such as the spectral imprints of the commodity of food in the THz region, enable the synthesis of THz sensing systems into real world.

This chapter analyzes the recent trends observed in THz technology and THz detection techniques and reports different issues that need to be addressed in order to apply THz sensing for food contamination detection. One application of contamination detection in fruit slices was provided through the transmission response in the THz frequency region.

5.1   THz components and application systems

5.1.1   Components of THz systems

The THz systems are mainly formed by three main modules such as THz sources, components, and detectors [40], Figure 5.1 depicts a typical THz configuration. The terahertz source can produce a broad range of THz radiation [40,41]. Several

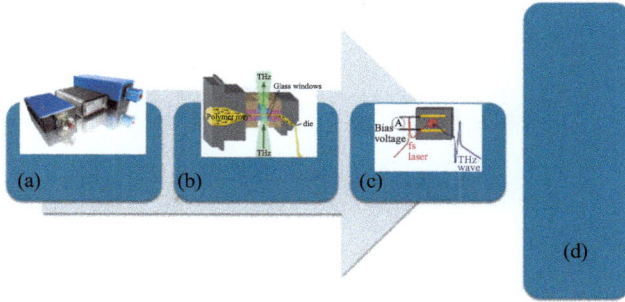

Figure 5.1 Typical terahertz configuration: (a) sources; (b) components; (c) detectors; and (d) applications [92]

new THz source techniques have been brought forth during the past few decades [42–47]. The THz components including mirrors, lenses, and polarizers manipulate the specific radiation. Modern innovations in THz mirror technology, which include semiconductor [48], hybrid mirrors [49], and the tunable mirrors, are based on photonic crystals [50]. Terahertz lenses are typically made of plastics such as polyethylene or TPX. Recently, the significant innovations allow the rapid production of a large number of lenses which include Fresnel zone plates, plasmonic resonances, variable focal length lenses, 3D printed diffractive lenses, and even THz lenses made of paper [51–56]. The polarizers are one of the significant components in THz imaging, data transmission, and spectroscopy. Recently, reconfigurable polarizers and carbon nanotube fiber polarizers have been reported [57,58]. The THz detectors that can measure the THz radiation upon reception play a significant role in many areas including astrophysics, chemical, biological, and explosive detection, imaging, astronomy applications, and so on [59,60].

5.1.2 Applications of THz systems

The broad applications of the THz radiation can be categorized into four main groups—sensing, imaging, spectroscopy, and communication—and can also be applied in preventive health care to quality control, surgery, and nondestructive evaluation (NDE) [61–64] as shown in Figure 5.2.

The potential of THz technologies in the food detection fields is listed in Table 5.1, which presents various other facets including the identification of foreign bodies, detection of pesticide and antibiotic residues, characterization of edible oil, discrimination of transgenic crops, and so on. For example, the identification of undesired and potentially unhealthy organisms in food is excessively important in the food industry [65]. With regard to the advantages of nondestructive, noninvasive, and nonionizing nature, the THz technologies are termed as the alternative methods for sensing food safety control and contamination detection.

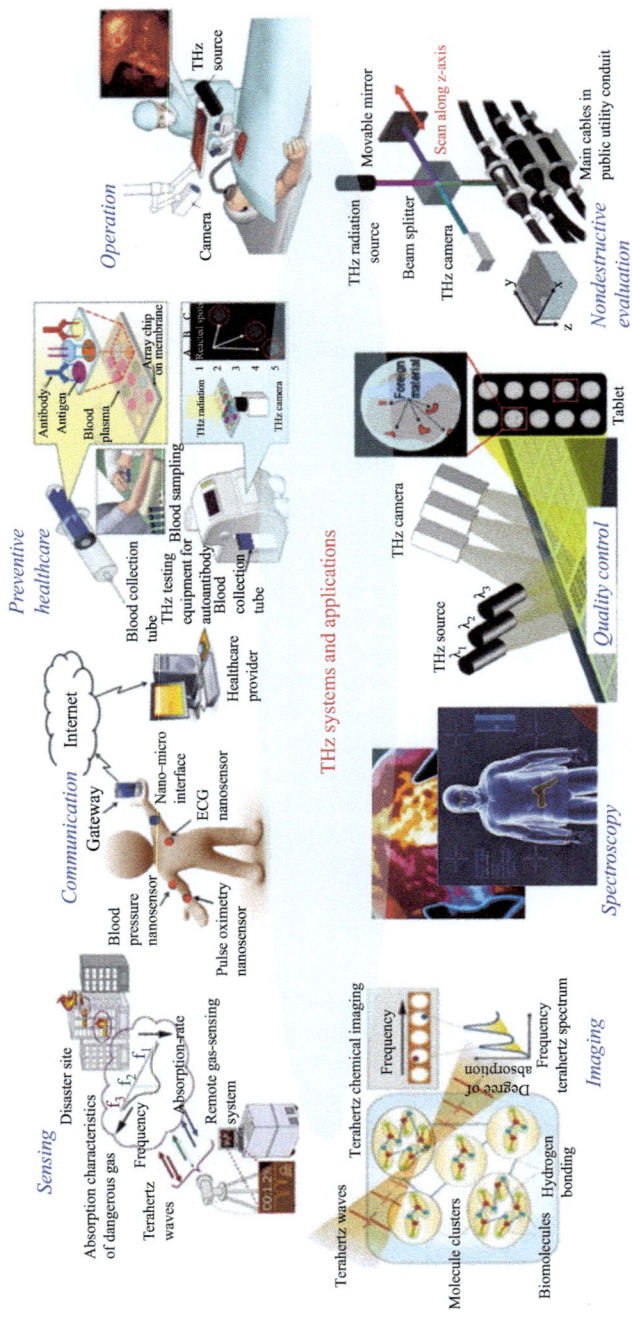

Figure 5.2 Envisioned applications for THz radiation [92]

Table 5.1   The applications of terahertz technology for food detection

Methods	Detected materials
THz-TDS	Melamine detection in foods [67]
	Detection of a carbamate insecticide in food matrices [93]
	Moisture content in wheat grain [94] and wafers [95]
	Pesticide detection in food powders [96]
	Antibiotic detection in food and feed matrices [97]
	Discrimination of transgenic crops [98,99]
	Identification of adulterated dairy product [100]
	Detection of harmful chemical residues in honey [101]
	Characterization of optical properties of vegetable oil [102]
THz-FDS[a]	Recognition of CCH[b] and TCH[c] in soil, chicken, and rice [103]
THz imaging	Metallic and nonmetallic foreign bodies in chocolate [104]

[a]THz-FDS: terahertz frequency-domain spectroscopy.
[b]CCH: chlortetracycline hydrochloride.
[c]TCH: tetracycline hydrochloride.

5.2   THz detecting methods and their enabling technologies

The field of THz technology has changed rapidly in the past few years. The research work on terahertz technology has greatly promoted the generation and development of many new technologies such as THz-TDS, THz imaging, and THz spectrum analysis with vector network analyzer (VNA). These technologies have been widely studied and applied in various fields.

5.2.1   THz time-domain spectroscopy

The THz-TDS can be configured in three modes, including transmission, reception, and attenuated total reflection (ATR) modes. A typical setup of the THz-TDS system and the three modes are shown in Figure 5.3. Compared with transmission and reflection modes, the ATR THz-TDS is highly sensitive and suitable for measuring high moisture samples. The TDS system has the potential to acquire direct measurement of the field amplitude information and phase information as a function of time by sweeping out the transient field of the THz pulse. The terahertz frequency range can be accessed by applying Fourier transformation on time-domain data. As the effective method of identifying the material spectra information in the THz region using ultrafast laser beam technology, THz-TDS has comparatively a high signal-to-noise ratio due to the broader frequency range and can be used to determine the nature of the material (or composition) and profound structural changes. For example, the molecular classification and the associated imprint peak of three plant growth regulators were detected in [66]. The melamine in food products was recognized and shown the characteristic absorption peaks in different THz frequencies [67]. The research results showed that the wave

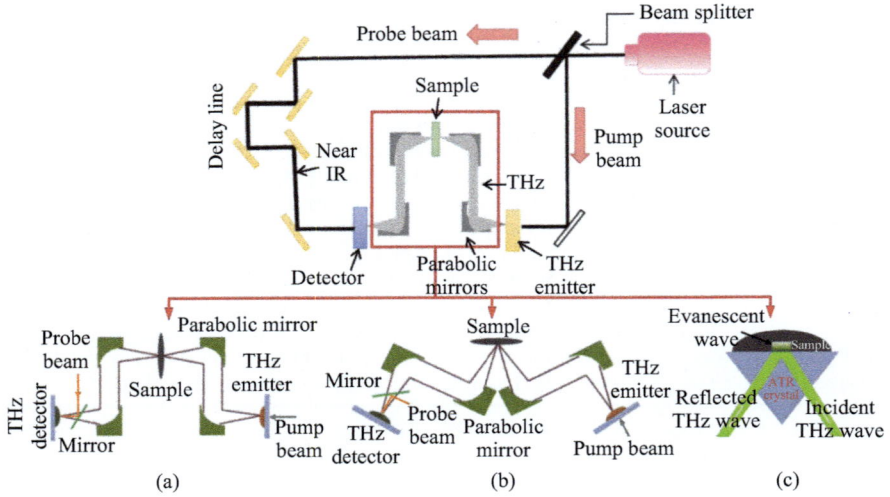

Figure 5.3   *Schematic of a typical THz time-domain spectroscopy system:*
(a) transmission mode; (b) reflection mode; and (c) ATR mode [92]

propagation velocity was lower in the contamination region compared to the pure material [68].

5.2.2   Terahertz imaging

The terahertz imaging is an emerging and significant NDE technique used for safety checks, quality control, and even fanciful applications, such as using terahertz image technology to count almond chocolates in bars [69]. One of the great advantages of terahertz in the field of image processing is that terahertz radiation has a unique ability to penetrate ordinary packaging materials, so it can provide spectral information of internal materials.

5.2.3   THz spectrum analysis with VNA

The VNA operates in the frequency domain, which can be applied to measure the amplitude and phase of an incident signal interacting with a device-under-test [70]. Recently, with the frequency extender heads fitted with standardized rectangular metal waveguides, THz spectrum analysis with VNA extended to higher frequencies can be employed to measure the complex scattering parameters of THz radiation incident on a test device, which are called S-parameter such as reflection and transmission measurements (i.e., S_{11}, S_{21}, S_{12}, and S_{22}). The typical VNA system for THz measurements will be introduced in Section 5.4.2. Presently, as commercial THz VNAs can operate at frequencies up to 1.1 THz [70], they can be applied to measure the material characterization, in which they can configure free-space measurements similar to TDS [71,72].

5.3   Challenges in food contamination detection

The terahertz nondestructive detection technology in food contamination detection is emerging as a new area of study. With unique superiority, THz technology has attracted many researchers in this field. This field has made great progress; however, it is still at an initial stage, and there are still many challenges that need to be addressed. Some of the most important challenges are given as follows.

5.3.1   Distraction and absorption of water content

Terahertz nondestructive detection for food contamination detection is limited by water content (WC), thanks to the firm absorption of THz radiation by water. Therefore, a major challenge is the overwhelming fading of THz radiation by water molecules. For example, in food detection, THz technology is not appropriate for detecting high humidity products with a thickness greater than 1 mm. The inability of THz technology to detect the biomolecular interactions in solution is the major hurdle facing terahertz further applications.

5.3.2   Low penetration depth of THz radiation

Another challenge of terahertz technology is the low penetration depth of terahertz radiation, especially measuring the liquid and meat products. This problem can be addressed using the PEAs, the use of graphene composite, and strengthening the intensity of the terahertz source.

5.3.3   Scattering effects

Scattering effects are also a common problem in terahertz time-domain spectroscopy transmission measurements of solid-state samples, especially for irregular granular samples. To reduce or remove the scattering effects, plasma and metamaterial can be applied [73]. In addition, some better algorithms can also be used, such as wavelet transform and the Monte Carlo method, to extract spectral data to exclude or reduce the scattering effects [74].

5.4   State-of-the-art method for fruit spoilage detection

Due to the safety of THz radiation to tissues and biomolecules of food and any live cells, the THz technique has become a novel and powerful nondestructive technologies to be explored for monitoring the quality of food and recognizing the inner intermolecular characteristics of biological materials in the THz range. The antibiotic and harmful pesticide in food could be detected because of the intrinsic resonance properties of chemicals and biomolecules in the THz radiation region [75,76]. Unlike some existing nuclear magnetic resonance and near-infrared (NIR) techniques [77,78], the THz technology can obtain more detailed internal characteristics of fruits due to its high resolution and sensitivity at molecular vibration. Meanwhile, the THz sensing can be employed to detect the refractive index, absorption coefficient, and the dielectric properties of the samples. It has been proved that the dielectric properties of fruits and vegetables have some connection with moisture content (MC) or water activity, which is the main factor that causes the changes of electrical characteristics because of the different inner substances present in fruits and vegetables [79,80]. Consequently, the quality of fruit is mainly dominated by the MC value.

Even though many policies and efforts have been made to prevent food contamination, it is more important and challenging to inspect the quality of food products and evaluate the contents of food materials. Specifically, the safe methods for the biomaterials and the agricultural products are of utmost importance, which are effective to identify contaminated food, to detect toxic substances, such as pesticide residues, additives, antibiotics, and pathogens in foodstuffs. Although many technologies have been used for qualitative and quantitative analysis [81] in a wide range of food products, such as NIR spectroscopy imaging, bioassays, and molecular imprint-based sensors [82], the nonionizing and nondestructive THz radiation is considered the novel, accurate, and economical way for analyzing the food compositions.

5.4.1 Detection of moisture content

Moisture content (MC), or WC, is the quantity of water contained in a material, such as seed and grain, food and beverage, agricultural and soil, pulp and paper products, and petroleum products. The THz radiation has the potential advantages for sensing and imaging of the moisture map, nondestructively assessing hydration levels in materials. The first application of THz imaging and sensing for detecting the WC is to monitor and evaluate the moisture level of plant leaves, which can provide valuable information to farmers and scientists regarding plant drought stress and irrigation management [83]. The theoretical basis for detecting the WC in plants can be conducted to establish prediction models of MC based on the processed THz transmission and absorption spectra [84]. The meaningful information attained from the water detection of the living plant leaves can be useful to analyze the existence of any pesticides in leaves.

Depending on the different absorption coefficients and refraction index of moisture on the thin surface layers of fruits, the THz wave can be used to evaluate the internal quality of fruit by nondestructive detection. The THz reflectivity from the pressed regions of the fruits revealed the decreased tendency due to the damaged area losing the moisture [85]. Through the detection of the refractive index and the absorption coefficient in the regression model, the soluble solid content, which is an important index for fruit quality, was determined in apple products using the THz-TDS technique [86]. The potential of dielectric spectroscopy related to temperature, soluble solid content, and MC with the fruit gained by THz technology can be applied for characterizing the quality factors of fruits or vegetables [87].

As the lower MC inhibits the activities of food-spoilage and food-poisoning microorganisms, the low-moisture foods, or dried foods, are more easily preserved.

However, it is significant to detect and monitor the MC of dried food for avoiding mold growth when the WC exceeds the normal level [88]. The advantages of THz technology are that the THz absorbance of water is higher than the other constituents, and no significant heating is produced in the sample due to the minimal THz power [83].

5.4.2 Materials and methods

This section provides a simple method of employing THz technology to monitor the variations of the MC value of apple and pear slices and derives the absorption coefficient by employing the transmission response from measurements of the VNA with the Swissto12 system in the THz frequency range of 0.75–1.1 THz. These preliminary results provided significant information to study the more detailed characteristics of substances existing in the inner fruits through the more dielectric properties, including permittivity, permeability, and refractive index in the THz frequency region.

5.4.2.1 Experimental setup

The experimental setup consists of three parts—VNA, the Material Characterization Kit (MCK) Swissto12 waveguide system, and two waveguide extender ports in the frequency ranging from 0.75 to 1.1 THz, as shown in Figure 5.4(a) and (b). Prior to taking any measurements, an appropriate calibration is required with a two-port Short-Open-Load-Thru criterion to diminish the losses existed in the system and the errors occurred while taking measurements. Subsequently, the scattering parameters (S-parameter) of samples including the transmission (S12, S21) and reflection (S11, S22) responses can be determined based on VNA in the frequency range of 0.75–1.1 THz.

5.4.2.2 Sample details

Two fruit slices, including apple and pear, were taken as samples with the average thickness between 2.3 and 3.5 mm, as shown in Figure 5.5 for pear slices. All the sample slices were analyzed and preserved carefully under the ambient temperature of 18 °C $\pm$ 0.1 °C and a humidity of 30% $\pm$ 2%. The thickness of each slice was measured using a Vernier caliper at three different locations to ensure similar surface area and distribution all over the whole slice and satisfy the threshold range of MCK at the beginning of the experiment. Moreover, the weight of each slice was collected before taking the measurement employing a digital scale with an accuracy of 0.1 mg to be used to calculate the MC value.

Before collecting observations, meanwhile, the weight of each slice was also measured using a digital scale with an accuracy of 0.1 mg to monitor the changes of MC in samples. The weight of each sample was converted into the MC value with the passing time as

$$Value_{MC} = \frac{W_{sample} - W_{dry}}{W_{sample}} \times 100\% \tag{5.1}$$

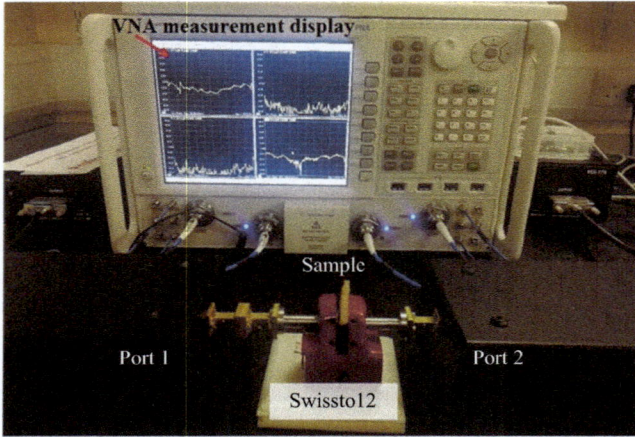

(a)

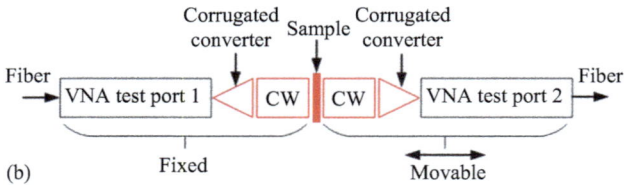

(b)

Figure 5.4   Experimental setup of THz system for measuring the transmission response with Swissto12: (a) experimental system and (b) schematic diagram

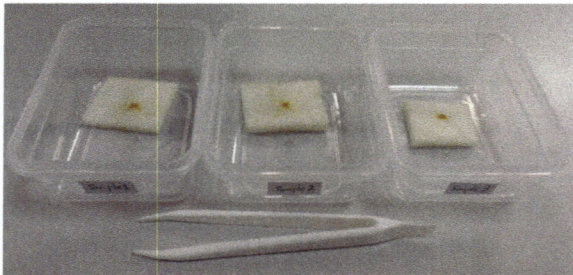

Figure 5.5   Three pear slices

where $Value_{MC}$ is the MC value, W_{sample} indicates the weight of slice at the measuring time, and W_{dry} denotes the weight of the slice dried out completely. The variation rate and the range of MC of apple and pear slices are shown in Figure 5.6. The MC patterns of different slices, as shown in Figure 5.6, indicate the decaying patterns of the freshness of the pear and apple slices which were exposed to air as the days progressed.

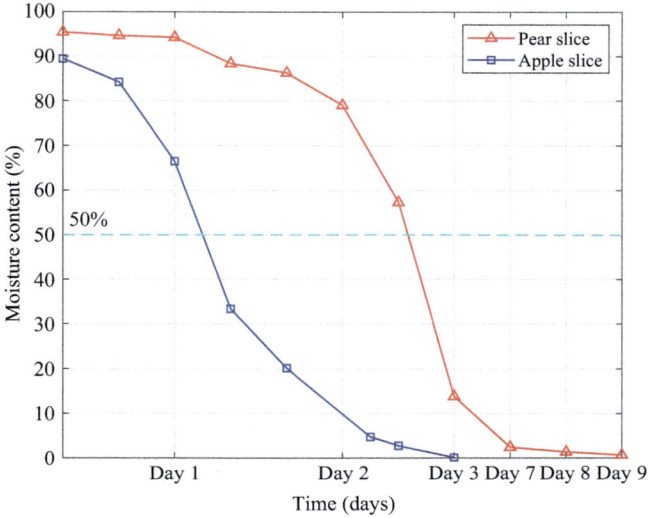

Figure 5.6   Moisture contents of two slices with days

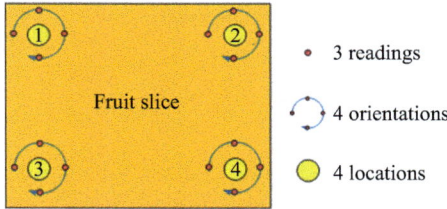

Figure 5.7   4 × 4 × 3 measurements were collected for each slice at each time

5.4.3   Measurement results

In order to obtain the average measurements, each slice was taken at four different locations, and each location was measured for orientation with three readings each time in consecutive days until the sample slices were dried out fully, as shown in Figure 5.7.

The average transmission response of three times measurements of each angle was calculated by using

$$\widetilde{S}_{21(k)} = \frac{\sum_{i=1}^{3} S_{21(k)}^{(i)}}{3} \tag{5.2}$$

where $S_{21(k)}^{(i)}$ is the measurement of the i-th time of the k-th sample and $\widetilde{S}_{21(k)}$ is the average transmission response of three-times measurements of the k-th sample.

5.4.3.1   Transmission response

There are eight data sets collected from VNA, $S_{(k)}^{(n)}(f_i)$, which are for two different slices ($k = 1$, 2 numbered for apple slices and pear slices, respectively) in continuous days ($n = 1, 2, 3, ...$) and changed with frequency f_i ($i = 0:200$) in the range of 0.75–1.1 THz. As an example, Figures 5.8 and 5.9 show one observation of the transmission response (S21) of apple and pear slices with the passing days, respectively. The raw observations of the fresh fruit slices show the different transmission responses in the THz region with the passing days due to the evaporation of MC.

The transmission response of pear and apple slices, shown in Figure 5.7(a) and (b), illustrates the THz characteristics of two fruits in the same THz region. A distinct variation can be observed in transmission response of the pear slice on day 3 to day 9 as shown in the embedded fitting curves; however, the difference was not very noticeable on day 1 to day 3 due to a high MC value. Compared to the pear slices, the significant difference was observed on day 1 to day 3 for the apple slice as shown in Figure 5.7(a), even on the same day between two times measurements. Observed the trend of MC as shown in Figure 5.7(b), the difference of the water loss from day 1 to day 3 was 60.88% for the pear slice and 49.79% for the apple slice. Nevertheless, the difference of the MC value was very significant, which was only 15.26% for the pear slice and 61.38% for the apple slice.

5.4.3.2   Path-loss response of fruit slices

The path loss of apple and pear slices obtained over the course of passing days is shown in Figures 5.10 and 5.11, respectively. From the embedded fitting curves, the samples showed a clear difference between transmission and absorption with

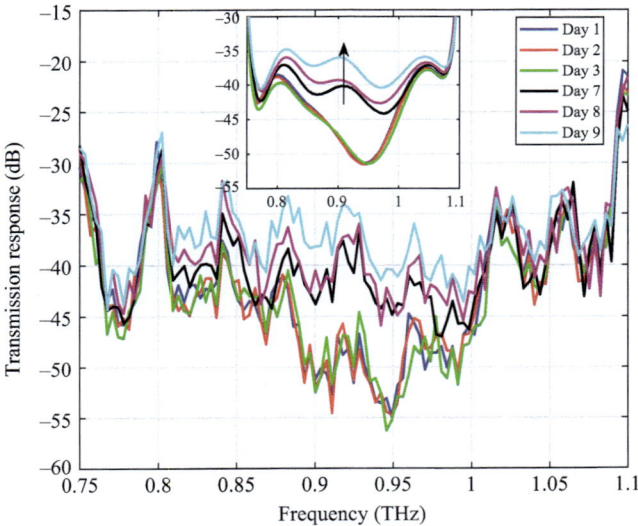

Figure 5.8   Transmission response of apple slice

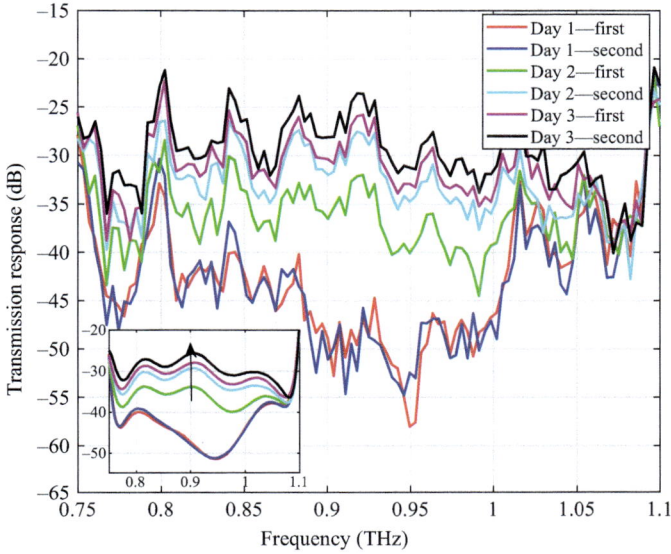

Figure 5.9   Transmission response of pear slice

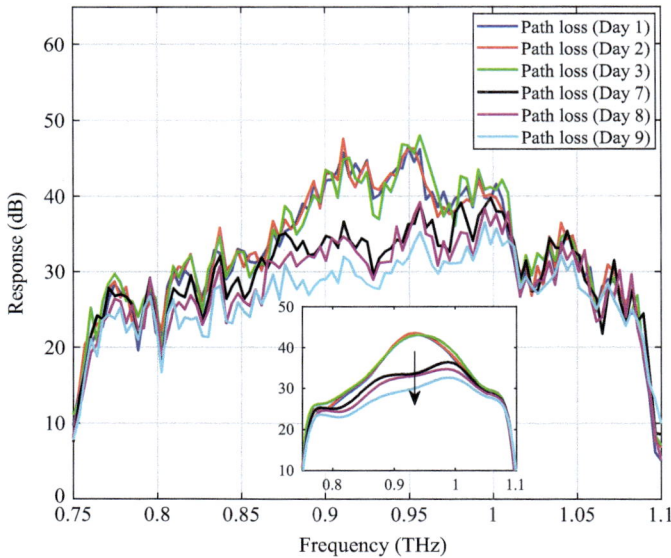

Figure 5.10   Path-loss response of apple slice

the variation of MC, as shown in Figure 5.6, in the THz frequency range from 0.85 to 1.0 THz. Meanwhile, observed the surface of the pear and apple slices, the appearance of pear slice was stickier than that of the apple slice with the passing days, which indicated that different internal characterizations of fruits related to the MC.

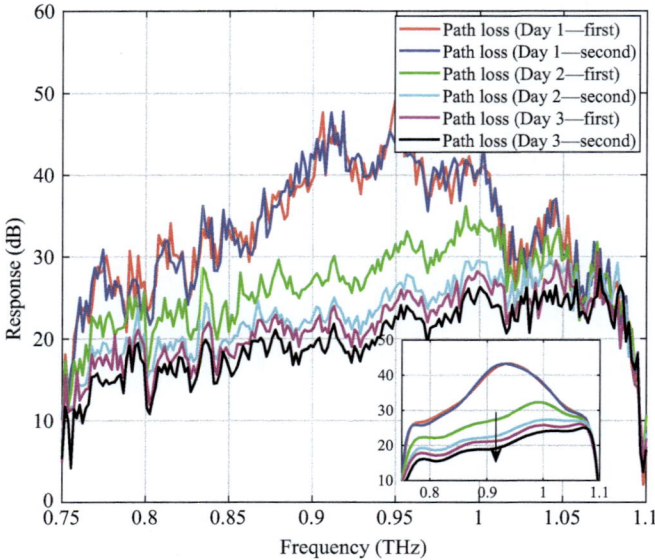

Figure 5.11   Path-loss response of pear slice

The results of the transmission response and the path-loss response of apple and pear slices indicated that the different substances in fruits will merge the different THz characteristics in the THz region due to the loss of the moisture, which is one of the important factors to detect the quality of the fruits. This method in this paper can be extended to analyze and control the quality of fruits and vegetables in terms of identifying the MC.

5.4.3.3   Correlation of absorption coefficient with the transmission response

To calculate the absorption coefficient of the sample, $\alpha_s(\omega)$, the reflection coefficient, $r_s(\omega)$, and transmission coefficient, $t_s(\omega)$, were obtained from the measured S-parameters, respectively, as [89]:

$$r_s(\omega) = X \pm \sqrt{X^2 - 1} \tag{5.3}$$

$$t_s(\omega) = \frac{S_{11} + S_{21} - r_s(\omega)}{1 - (S_{11} + S_{21})r_s(\omega)} \tag{5.4}$$

Here, the reflection coefficient should comply with $|r_s(\omega)| \leq 1$ for passive samples. ω is the angular frequency of the THz waves. The middle parameter, X, can be derived from:

$$X = \frac{1 + (S_{11} - S_{21})(S_{11} + S_{21})}{2S_{11}} \tag{5.5}$$

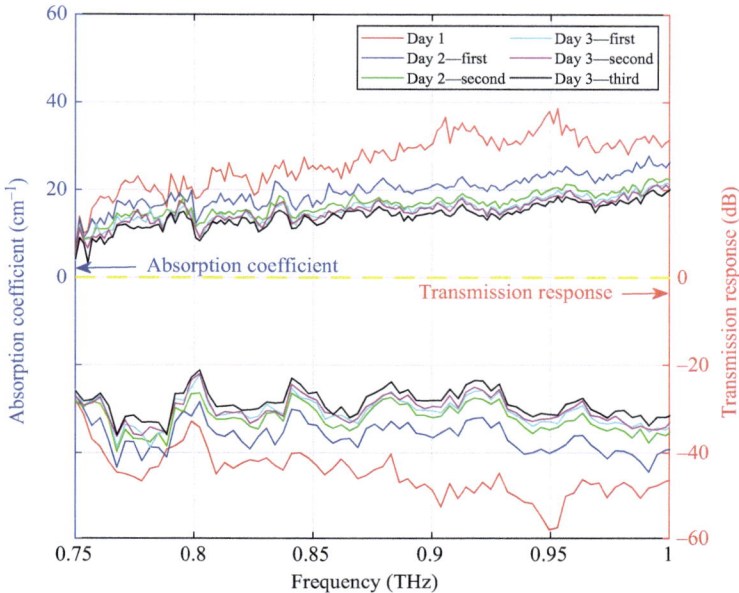

Figure 5.12   The relation of absorption coefficient of apple slice with transmission response from 0.75 to 1.0 THz with days

Moreover, with $t_s(\omega) = \exp\{j\omega dn_s(\omega)\}$, the complex refractive index, $n_s(\omega)$, can be obtained. d is the thickness of the sample. Finally, the absorption coefficient, $\alpha_s(\omega)$, can be extracted from [90]:

$$\alpha_s(\omega) = \frac{2\omega \cdot \mathrm{Imag}\{n_s(\omega)\}}{c} \tag{5.6}$$

where $\mathrm{Imag}\{*\}$ indicates the imaginary part and c is the velocity of light.

The transmission response obtained directly from VNA appeared the attenuation of slices in the EM wave region. Here, the absorption coefficient and the transmission response of apple and pear slices against the THz frequency are shown in Figures 5.12 and 5.13, respectively. The top half of the figure shows the absorption coefficient corresponding to the left y-axis with the unit of cm^{-1}, and the bottom represents the transmission response corresponding to the right y-axis with the unit of dB. The samples showed a clear increase in the trend of the absorption characteristic with the frequency in the range of 0.75–1.0 THz; nevertheless, the transmission response of the samples was decreased with the frequency. Comparing Figure 5.12 with Figure 5.13, it is obviously illustrated that the relationship between the absorption coefficient and transmission response was very clear when the MC value was lower than 50% referred to Figure 5.6.

Due to the different texture and inner substances of pear and apple, especially the loss rate of the water with the passing days, the apple slice showed the

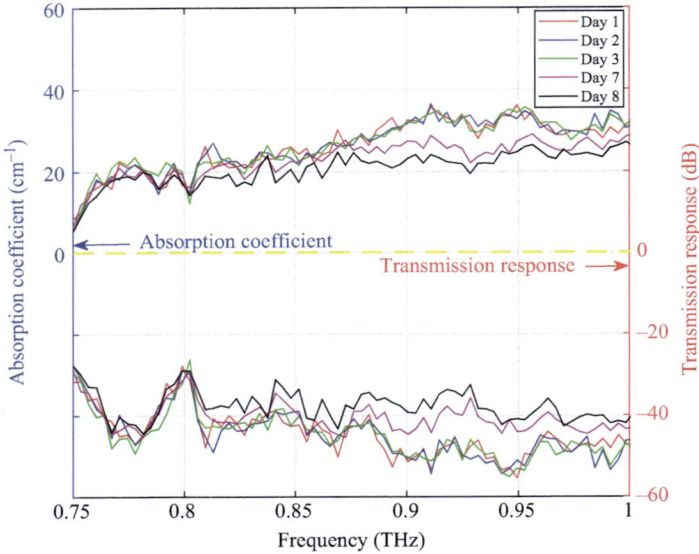

Figure 5.13   The relation of absorption coefficient of pear slice with transmission response from 0.75 to 1.0 THz with days

significant difference in absorption characteristic and transmission response on day 1 to day 3. Nevertheless, the pear slice illustrated a distinct variation in the absorption coefficient and transmission response until the MC value was declined to lower than 50% from day 3 to day 8. In Figures 5.12 and 5.13, the strong absorption peak can be observed clearly when the MC value is larger than 50% due to more water in the samples. That indicates that the absorption of the water to the THz radiation played a decisive role in this case. With the passing days, the loss of the water was more and more, and the difference in the absorption characteristic of the apple and pear slices was more distinguished in the THz region, which can be employed to determine the presence of pesticides or more intermolecular information of fruits and further realize the quality and security control.

5.4.3.4   Correlation of absorption coefficient with MC

For the investigated slices of this section, the relationship between the absorption coefficients and the MC value at the frequency of 0.95 THz is shown in Figure 5.14. It depicts the absorption of different slices as a function of the MC with the passing days, which also implies that the influence of the variation of the concentration of inner constituents in the fruits is very less in the case of neglecting the scattering losses [91].

Observed the trends of the curves, the blue fitting for pear slice and the red fitting for apple slice, as shown in Figure 5.14, the percentage of the reduced absorption from day 1 to day 3 was around 57% for the apple slice, while it was only about 28.6% from day 1 to day 8 for the pear slice. Due to the dominant position of MC, the related dielectric properties including the permittivity and the

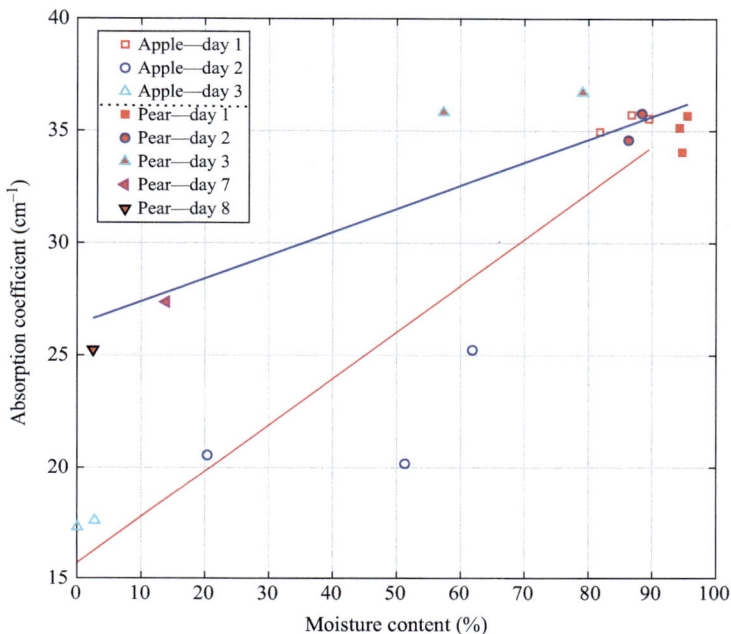

Figure 5.14 Absorption coefficient against moisture content of slices with days

permeability of samples can be further estimated to determine the chemical composition and texture structure of the different fruits for nondestructive quality control in the future research.

5.5 Conclusion

This chapter mainly focuses on various sensing technologies that have been employed to detect food and water contamination. It has been found that these conventional sensing technologies appear to be unfeasible and impractical to meet with the challenging growth of population. In this aspect, THz sensing is discussed in detail and deemed to be more effective due to its strong penetration feature, high resolution, and sensitivity to monitor the molecular changes in fruits. This chapter also introduces a novel technique of fruits contamination detection by monitoring MC and observe the transmission and path loss response of fruits. It also investigates an important parameter such as the absorption coefficient and shows some significant results and correlation of MC with transmission response and absorption coefficient. Upon close analysis, these results give meaningful information about the composites present in fruits such as carbohydrates and proteins. Toward the end, this chapter emphasizes on the advancement and development of terahertz technology applications and found that the THz sensing is a promising candidate and has a potential to change a paradigm in the plant science sector.

References

[1]   D. M. Mittleman, Perspective: Terahertz Science and Technology, Journal of Applied Physics, Vol. 122(23), 2017.

[2]   P. R. Smith, D. H. Auston, and M. C. Nuss, Subpicosecond Photoconducting Dipole Antennas, IEEE Journal of Quantum Electronics, Vol. 24(2), 1988.

[3]   J. Hebling, K.-L. Yeh, M. C. Hoffmann, B. Bartal, and K. A. Nelson, Generation of High-power Terahertz Pulses by Tilted-Pulse-Front Excitation and Their Application Possibilities, Journal of Optical Society of America B, Vol. 25(7), 2008, pp. B6–B19.

[4]   C. Vicaro, B. Monoszlai, and C. P. Hauri, GV/m Gingle-Cycle Terahertz Fields from a Laser-Driven Large-Size Partitioned Organic Crystal, Physical Review Letters, Vol. 112, 2014.

[5]   J. Dai, J. Liu, and X.-C. Zhang, Terahertz Wave Air Photonics: Terahertz Wave Generation and Detection With Laser-Induced Gas Plasma, IEEE Journal of Selected Topics in Quantum Electronics, Vol. 17(1), 2011.

[6]   M. C. Hoffmann, and J. A. Fulop, Intense Ultrashort Terahertz Pulses: Generation and Applications, Journal of Physics D: Applied Physics, Vol. 44, 2011, 083001.

[7]   R. M. Woodward, B. E. Cole, V. P. Wallance, *et al.*, Terahertz Pulse Imaging in Reflection Geometry of Human Skin Cancer and Skin Tissue, Physics in Medicine and Biology, Vol. 47, 2002, pp. 3853–3863.

[8]   S. Y. Huang, Y. X. J. Wang, D. K. W. Yeung, A. T. Ahuja, Y.-T. Zhang, and E. Pickwell-MacPherson, Tissue Characterization Using Terahertz Pulsed Imaging in Reflection Geometry, Physics in Medicine and Biology, Vol. 54, 2009, pp. 149–160.

[9]   B. Heshmat, G. M. Andrews, O. A. Naranjo-Montoya, *et al.*, Terahertz Scattering and Water Absorption for Porosimetry, Optics Express, Vol. 25(22), 2017.

[10] H. Hintzsche, C. Jastrow, T. Kleine-Ostmann, U. Kärst, T. Schrader, and H. Stopper, Terahertz Electromagnetic Fields (0.106 THz) Do Not Induce Manifest Genomic Damage In Vitro, PLOS ONE, Vol. 7(9), 2012, e46397.

[11] H. Cheon, H.-J. Yang, and J.-H. Son, Toward Clinical Cancer Imaging Using Terahertz Spectroscopy, IEEE Journal of Selected Topics in Quantum Electronics, Vol. 23(4), 2017.

[12] Y.-C. Shen, Terahertz Pulsed Spectroscopy and Imaging for Pharmaceutical Applications: A Review, International Journal of Pharmaceutics, Vol. 417(1), 2011.

[13] G. Piro, K. Yang, G. Boggia, N. Chopra, L. A. Grieco, and A. Alomainy, Terahertz Communications in Human Tissues at the Nanoscale for Healthcare Applications, IEEE Transactions on Nanotechnology, Vol. 14(3), 2015.

[14] N. Chopra, K. Yang, Q. H. Abbasi, K. A. Qaraqe, M. Philpott, and A. Alomainy, THz Time-Domain Spectroscopy of Human Skin Tissue for In-Body Nanonetworks, IEEE Transactions on Terahertz Science and Technology, Vol. 6(6), 2016, pp. 803–809.

[15]   Q. H. Abbasi, H. El Sallabi, N. Chopra, K. Yang, K. A. Qaraqe, and A. Alomainy, Terahertz Channel Characterization Inside the Human Skin for Nano-Scale Body-Centric Networks, IEEE Transactions on Terahertz Science and Technology, Vol. 6(3), 2016.

[16]   Q. H. Abbasi, K. Yang, N. Chopra *et al.*, Nano-Communication for Biomedical Applications: A Review on the State-of-the-Art From Physical Layers to Novel Networking Concepts, IEEE Access, Vol. 4(28), 2016, pp. 3920–3935.

[17]   T. K. Ostmann, and T. Nagatsuma, A Review on Terahertz Communications Research, Journal of Infrared, Millimeter, and Terahertz Wave, Vol. 32(2), 2011, pp. 143–171.

[18]   M. O. Iqbal, M. M. Ur Rahman, M. A. Imran, A. Alomainy, and Q. H. Abbasi, Modulation Mode Detection and Classification for In Vivo Nano-Scale Communication Systems Operating in Terahertz Band, IEEE Transactions on NanoBioscience, Vol. 18(1), 2019, pp. 10–17.

[19]   H. Zhan, K. Zhao, R. Bao, and L. Xiao, Monitoring PM2.5 in the Atmosphere by Using Terahertz Time-Domain Spectroscopy, Journal of Infrared Millimeter and Terahertz Waves, Vol. 37(9), 2016, pp. 929–938.

[20]   V. Dworak, S. Augustin, and R. Gebbers, Application of Terahertz Radiation to Soil Measurements: Initial Results, Sensors, Vol. 11(10), 2011, pp. 9973–9988.

[21]   A. A. Gowen, C. O'Sullivan, and C. P. O'Donnell, Terahertz Time Domain Spectroscopy and Imaging: Emerging Techniques for Food Process Monitoring and Quality Control, Trends in Food Science & Technology, Vol. 25(1), 2012, pp. 40–46.

[22]   K. Wang, D.-W. Sun, and H. Pu, Emerging Non-destructive Terahertz Spectroscopic Imaging Technique: Principle and Applications in the Agri-Food Industry, Trends in Food Science & Technology, Vol. 67, 2017, pp. 93–105.

[23]   H.-B. Liu, H. Zhong, N. Karpowicz, Y. Chen, and X.-C. Zhang, Terahertz Spectroscopy and Imaging for Defense and Security Applications, Proceedings of the IEEE, Vol. 95(8), 2007.

[24]   V. Damian, P. C. Logofătu, and T. Vasile, 3D THz Hyperspectrum Applied in Security Check-in, Proc. SPIE 10010, Advanced Topics in Optoelectronics, Microelectronics, and Nanotechnologies VIII, 100100Y, 14 December 2016.

[25]   J. Lu, Y. Zhang, H. Y. Hwang, B. K. Ofori-Okai, Fleischer, and K. A. Nelson, Nonlinear two-dimensional terahertz photon echo and rotational spectroscopy in the gas phase, Proceedings of the National Academy of Sciences USA, Vol. 113(42), 2016.

[26]   I. A. Finneran, R. Welsch, M. A. Allodi, T. F. Miller III, and G. A. Blake, Coherent Two-Dimensional Terahertz-Terahertz-Raman Spectroscopy, Proceedings of the National Academy of Sciences USA, Vol. 113(25), 2016, pp. 6857–6861.

[27]   J. Lu, X. Li, H. Y. Hwang, *et al.*, Coherent Two-Dimensional Terahertz Magnetic Resonance Spectroscopy of Collective Spin Waves, Physical Review Letters, Vol. 118, 2017.

[28]   S. K. Mathanker, P. R. Weckler, and N. Wang, Terahertz (THz) Applications in Food and Agriculture: A Review, Transactions of the ASABE, Vol. 56(1), 2013.

[29]   M. Heyden, J. Sun, S. Funkner, *et al.*, Dissecting the THz Spectrum of Liquid Water From First Principles via Correlations in Time and Space, Proceedings of the National Academy of Sciences USA, Vol. 107(27), 2010, pp. 12068–12073.

[30]   Y. Zhan Ke, Z. Hongjian, and Y. Yibin, Research Progress of Terahertz Wave Technology in Quality Measurement of Food and Agricultural Products, Spectroscopy and Spectral Analysis, Vol. 27, 2007.

[31]   Y. Ogawa, S. Hayashi, C. Otani, and K. Kawase, Terahertz Sensing for Ensuring the Safety and Security, PIERS Online, Vol. 4(3), 2008.

[32]   T. Suzuki, Y. Ogawa, and N. Kondo, Characterization of Pesticide Residue, Cis-permethrin by Terahertz Spectroscopy, Engineering in Agriculture Environment and Food, Vol. 4(4), 2011, pp. 90–94.

[33]   Z. Xiao-li, and L. Jiu-sheng, Diagnostic Techniques of Talc Powder in Flour Based on the THz Spectroscopy, Journal of Physics: Conference Series, Vol. 276(1), 012234.

[34]   Y. Cui, M. Kaijun, X. Wang, Y. Zhang, and C. Zhang, Measurement of Mixtures of Melamine Using THz Ray, Proc. SPIE, Vol. 7385.

[35]   L. Xie, Y. Yao, and Y. Ying, The Application of Terahertz Spectroscopy to Protein Detection: A Review, Applied Spectroscopy Reviews, Vol. 49, 2014, pp. 448–461.

[36]   L. Yang, H. Sun, S. Weng, *et al.*, Terahertz Absorption Spectra of Some Saccharides and Their Metal Complexes, Spectochimica Acta Part A: Molecular and Biomolecular Spectroscopy, Vol. 69(1), 2008, pp. 160–166.

[37]   J. F. Federici, B. Schulkin, F. Huang, *et al.*, THz Imaging and Sensing for Security Applications – Explosives, Weapons, and Drugs, Semiconductor Science and Technology, Vol. 20(7), 2005, pp. S266–S280.

[38]   S. Galoda, and G. Singh, Fighting Terrorism with Terahertz, IEEE Potentials, Vol. 26(6), 2007, pp. 24–29.

[39]   J. Xu, K. W. Plaxco, and S. J. Allen, Collective Dynamics of Lysozyme in Water: Terahertz Absorption Spectroscopy and Comparison with Theory, Journal of Physical Chemistry B, Vol. 110(47), 2006, pp. 24255–24259.

[40]   R. A. Lewis, A Review of Terahertz Sources, Journal of Physics D: Applied Physics, Vol. 47, 2014, 374001.

[41]   W. Ghann, and J. Uddin, Terahertz (THz) Spectroscopy: A Cutting-Edge Technology, Terahertz Spectroscopy, IntechOpen, 13 March 2017.

[42]   R. Bogue, Sensing with Terahertz Radiation: A Review of Recent Progress, Sensor Review, Vol. 38(2), 2018, pp. 216–222.

[43]   K. A. Fedorova, A. Gorodetsky, and E. U. Rafailov, Compact All-Quantum-Dot-Based Tunable THz Laser Source, IEEE Journal of Selected Topics in Quantum Electronics, Vol. 23(4), 2017, pp. 1–5.

[44]   B. Wu, Z. Zhang, L. Cao, Q. Fu, and Y. Xiong, Electro-optic Sampling of Optical Pulses and Electron Bunches for a Compact THz-FEL Source, Infrared Physics & Technology, Vol. 92, 2018, pp. 287–294.

[45]   S. Liu, High Peak Power THz Source for Ultrafast Electron Diffraction, AIP Advances, Vol. 8(1), 2018, 015029.

[46]   M. J. Nasse, M. Schuh, S. Naknaimueang, *et al.*, FLUTE: A Versatile LINAC-based THz Source, Review of Scientific Instruments, Vol. 84(2), 2013.

[47]   I. Morohashi, Y. Irimajiri, M. Kumagai, *et al.*, Terahertz source with broad frequency tunability up to 3.8 THz using MZ-FCG for frequency reference in phase-locking of THz source devices, IEEE International Topical Meeting on Microwave Photonics (MWP), 2016.

[48]   H. B. Ye, Y. H. Zhang, and W. Z. Shen, Carrier Transport and Optical Properties in GaAs Far-Infrared/Terahertz Mirror Structures, Thin Solid Films, Vol. 514, 2006, pp. 310–315.

[49]   N. Krumbholz, K. Gerlach, F. Rutz, and M. Koch, Omnidirectional Terahertz Mirrors: A Key Element for Future Terahertz Communication Systems, Applied Physics Letters, Vol. 88(20), 2006, 202905.

[50]   H. Zhang, X.-F. Xu, F. Fan, and S.-J. Chang, Study on a Tunable Narrow-band Filter Based on Magnetic Defects in Photonic Crystal in the Terahertz Region, Optical Engineering, Vol. 54(4), 2015, 047104.

[51]   B. Scherger, M. Scheller, C. Jansen, M. Koch, and K. Wiesauer, Terahertz lenses made by compression molding of micropowders, Applied Optics, Vol. 50, 2011, p. 2256.

[52]   H. D. Hristov, J. M. Rodriguez, and W. Grote, The Grooved-Dielectric Fresnel Zone Plate: An Effective Terahertz Lens and Antenna, Microwave and Optical Technology Letters, Vol. 54, 2012, p. 1343.

[53]   X.-Y. Jiang, J.-S. Ye, J.-W. He, *et al.*, An Ultrathin Terahertz Lens with Axial Long Focal Depth Based on Metasurfaces, Optics Express, Vol. 21, 2013, 30030.

[54]   B. Scherger, C. Jordens, and M. Koch, Variable-Focus Terahertz Lens, Optics Express, Vol. 19, 2011, p. 4528.

[55]   W. D. Furlan, V. Ferrando, J. A. Monsoriu, P. Zagrajek, E. Czerwińska, and M. Szustakowski, 3D Printed Diffractive Terahertz Lenses, Optics Letters, Vol. 41(8), 2016.

[56]   A. Siemion, A. Siemion, M. Makowski, *et al.*, Diffractive Paper Lens for Terahertz Optics, Optics Letters, Vol. 37, 2012, p. 4320.

[57]   L. J. Cheng, and L. Liu, Optical Modulation of Continuous Terahertz Waves Towards Cost-Effective Reconfigurable Quasi-Optical Terahertz Components, Optics Express, Vol. 21, 2013, 28657.

[58]   A. Zubair, D. E. Tsentalovich, C. C. Young, *et al.*, Carbon Nanotube Fiber Terahertz Polarizer, Applied Physics Letters, Vol. 108, 2016, 141107.

[59]   F. Sizov, and A. Rogalski, THz detectors, Progress in Quantum Electronics, Vol. 34(5), 2010, pp. 278–347.

[60]   A. Rogalski, and F. Sizov, Terahertz Detectors and Focal Plane Arrays, Opto-Electron, Rev., Vol. 19(3), 2011.

[61]   M. Razeghi, Q. Lu, S. Manna, D. Wu, and S. Slivken, Breakthroughs Bring THz Spectroscopy, Sensing Closer to Mainstream, Photonics Spectra, December 2016.

[62] J. Federici, and L. Moeller, Review of Terahertz and Subterahertz Wireless Communications, Journal of Applied Physics, Vol. 107(11), 2010.

[63] I. Hosako, and N. Oda, Terahertz Imaging for Detection or Diagnosis, SPIE Newsroom, June 8 2011.

[64] I. F. Akyildiz, J. M. Jornet, and C. Han, Terahertz Band: Next Frontier for Wireless Communications, Physical Communication, Vol. 12, 2014, pp. 16–32.

[65] I. Malhotra, K. R. Jha, and G. Singh, Terahertz Antenna Technology for Imaging Applications: a Technical Review, International Journal of Microwave and Wireless Technologies, Vol. 10(3), 2018, pp. 271–290.

[66] F. Qu, L. Lin, C. Cai, T. Dong, Y. He, and P. Nie, Molecular Characterization and Theoretical Calculation of Plant Growth Regulators Based on Terahertz Time-Domain Spectroscopy, Applied Sciences, Vol. 8, 2018, p. 420.

[67] S. H. Baek, H. B. Lim, and H. S. Chun, Detection of Melamine in Foods Using Terahertz Time-Domain Spectroscopy, Journal of Agricultural and Food Chemistry, Vol. 62(24), 2014, pp. 5403–5407.

[68] M. Mieloszyk, K. Majewska, and W. Ostachowicz, Composite Samples with Different Contaminations Analysed with THz Spectrometry, Proc. SPIE 10600, Health Monitoring of Structural and Biological Systems XII, 106001R, 27 March 2018.

[69] G. Ok, H. J. Shin, M.-C. Lim, and S.-W. Choi, Large-Scan-Area Sub-Terahertz Imaging System for Nondestructive Food Quality Inspection, Food Control, Vol. 96, 2018, pp. 383–389.

[70] M. Naftaly, G. C. Roland, D. A. Humphreys, and N. M. Ridler, Metrology State-of-the-Art and Challenges in Broadband Phase-Sensitive Terahertz Measurements, Proceedings of the IEEE, Vol. 105(6), 2017, pp. 1151–1165.

[71] B. Yang, X. Wang, Y. Zhang, and R. S. Donnan, Experimental Characterization of Hexaferrite Ceramics from 100 GHz to 1 THz Using Vector Network Analysis and Terahertz-Time Domain Spectroscopy, Journal of Applied Physics, Vol. 109, 2011, 033509.

[72] W. Sun, B. Yang, X. Wang, Y. Zhang, and R. S. Donnan, Accurate Determination of Terahertz Optical Constants by Vector Network Analyzer of Fabry–Perot Response, Optics Letters, Vol. 38, 2013, pp. 5438–5441.

[73] J. Ji, J. Jiang, J. Chen, F. Du, and P. Huang, Scattering Reduction of Perfectly Electric Conductive Cylinder by Coating Plasma and Metamaterial, Optik, Vol. 161, 2018, pp. 98–105.

[74] V. L. Malevich, G. V. Sinitsyn, G. B. Sochilin, and N. N. Rosanov, Manifestations of Radiation Scattering in the Method of Pulsed Terahertz Spectroscopy, Optics and Spectroscopy, Vol. 124(6), 2018, pp. 889–894.

[75] A. Redo-Sanchez, G. Salvatella, R. Galceran *et al.*, Assessment of Terahertz Spectroscopy to Detect Antibiotic Residues in Food and Feed Matrices, Analyst, Vol. 136(8), 2011, pp. 1733–1738. https://doi.org/10.1039/c0an01016b.

[76] Y. Hua, and H. Zhang, Qualitative and Quantitative Detection of Pesticides with Terahertz Time-Domain Spectroscopy, IEEE Transactions on Microwave Theory and Techniques, Vol. 58(7 part 2), 2010, pp. 2064–2070.

[77] A. D. C. Santos, F. A. Fonseca, L. M. Liao, G. B. Alcantara, and A. Barison, High-Resolution Magic Angle Spinning Nuclear Magnetic Resonance in Foodstuff Analysis, Trends in Analytical Chemistry, Vol. 73, 2015, pp. 10–18.

[78] R. Beghi, G. Giovanelli, C. Malegori, V. Giovenzana, and R. Guidetti, Testing of a VIS-NIR System for the Monitoring of Long-Term Apple Storage, Food and Bioprocess Technology, Vol. 7(7), 2014, pp. 2134–2143.

[79] O. Sipahioglu, and S. A. Barringer, Dielectric Properties of Vegetables and Fruits as a Function of Temperature, Ash, and Moisture Content, Food Science, Vol. 68(1), 2003, pp. 234–239.

[80] D. Khaled, N. Novas, J. Gazquez, R. Garcia, and F. Manzano-Agugliaro, Fruit and Vegetable Quality Assessment via Dielectric Sensing, Sensors, Vol. 15(7), 2015, pp. 15363–15397.

[81] Y. Liu, and Y. Ying, Use of FT-NIR Spectrometry in Non-invasive Measurements of Internal Quality of 'Fuji' Apples, Postharvest Biology and Technology, Vol. 37, 2005, pp. 65–71.

[82] M. Defernez, E. K. Kemsley, and R. H. Wilson, Use of Infrared Spectroscopy and Chemometrics for the Authentication of Fruit Purees, Journal of Agricultural and Food Chemistry, Vol. 43, 1995, pp. 109–113.

[83] J. F. Federici, Review of Moisture and Liquid Detection and Mapping using Terahertz Imaging, Journal of Infrared, Millimeter, and Terahertz Waves, Vol. 33(2), 2012, pp. 97–126.

[84] P. Nie, F. Qu, L. Lin *et al.*, Detection of Water Content in Rapeseed Leaves Using Terahertz Spectroscopy, Sensors, Vol. 17, 2017, p. 2830.

[85] Y. Ogawa, S. Hayashi, N. Kondo, K. Ninomiya, C. Otani, and K. Kawase, Feasibility on the Quality Evaluation of Agricultural Products with Terahertz Electromagnetic Wave, in 2006 ASABE Annual International Meeting, 2006, Portland, Oregon, 063050.

[86] Guohui Hao, Jianjun Liu, and Zhi Hong, Determination of Soluble Solids Content in Apple Products by Terahertz Time-Domain Spectroscopy, Proc. SPIE 8195, International Symposium on Photoelectronic Detection and Imaging: Terahertz Wave Technologies and Applications, 819510, 11 August 2011.

[87] D. El Khaled, N. Novas, J. A. Gazquez, R. M. Garcia, and F. M. Agugliaro, Fruit and Vegetable Quality Assessment via Dielectric Sensing, Sensors, Vol. 15, 2015, pp. 15363–15397.

[88] T. P. Labuza, and C. R. Hyman, Moisture Migration and Control in Multi-domain Foods, Trends in Food Science & Technology, Vol. 9(2), 1998, pp. 47–55.

[89] Luukkonen, Olli, Stanislav I. Maslovski, and Sergei A. Tretyakov, A Stepwise Nicolson–Ross–Weir-based Material Parameter Extraction Method, IEEE Antennas and Wireless Propagation Letters, Vol. 10, 2011, pp. 1295–1298.

[90] D.-K. Lee, J.-H. Kang, J.-S. Lee *et al.*, Highly Sensitive and Selective Sugar Detection by Terahertz Nano-antennas, Scientific Reports, Vol. 5, 2015, p. 15459.

[91] P. Parasoglou, E. P. J. Parrott, J. A. Zeitler *et al.*, Quantitative Water Content Measurements in Food Wafers Using Terahertz Radiation, IEEE Transactions on Terahertz Science and Technology, Vol. 3(4), 2010, pp. 176–182.

[92]   A. Ren, A. Zahid, D. Fan *et al.*, State-of-the-Art in Terahertz Sensing for Food and Water Security – A Comprehensive Review, Trends in Food Science & Technology, Vol. 85, 2019, pp. 241–251.

[93]   S. H. Baek, J. H. Kang, Y. H. Hwang, K. M. Ok, K. Kwak, and H. S. Chun, Detection of Methomyl, a Carbamate insecticide, in Food Matrices Using Terahertz Time-Domain Spectroscopy, Journal of Infrared, Millimeter, and Terahertz Waves, Vol. 37(5), 2016, pp. 486–497.

[94]   H. S. Chua, P. C. Upadhya, A. D. Haigh, J. Obradovic, and A. A. P. Gibson, Conference Digest of the 2004 Joint 29th International Conference on Terahertz Time-Domain Spectroscopy of Wheat Grain, Infrared and Millimeter Waves, 2004 and 12th International Conference on Terahertz Electronics.

[95]   P. Parasoglou, E. P. J. Parrott, J. A. Zeitler *et al.*, Quantitative Moisture Content Detection in Food Wafers, 34th International Conference on Infrared, Millimeter, and Terahertz Waves, Busan, 21–25 September 2009, pp. 1–2.

[96]   Y. Hua, and H. Zhang, Qualitative and Quantitative Detection of Pesticides With Terahertz Time-Domain Spectroscopy, IEEE Transactions on Microwave Theory and Techniques, Vol. 58(7), 2010, pp. 2064–2070.

[97]   A. R. Sanchez, G. Salvatella, R. Galceran *et al.*, Assessment of Terahertz Spectroscopy to Detect Antibiotic Residues in Food and Feed Matrices, Analyst, Vol. 136, 2011, pp. 1733–1738.

[98]   W. Liu, C. Liu, F. Chen, J. Yang, and L. Zheng, Discrimination of Transgenic Soybean Seeds by Terahertz Spectroscopy, Scientific Reports, Vol. 6(1), 2016, pp. 35799.

[99]   W. Xu, L. Xie, Z. Ye *et al.*, Discrimination of Transgenic Rice Containing the Cry1Ab Protein Using Terahertz Spectroscopy and Chemometrics, Scientific Reports, Vol. 5(1), 2015, pp. 11115.

[100]   J. Liu, Terahertz Spectroscopy and Chemometric Tools for Rapid Identification of Adulterated Dairy Product, Optical and Quantum Electronics, Vol. 49(1), 2017.

[101]   M. Massaouti, C. Daskalaki, A. Gorodetsky, A. D. Koulouklidis, and S. Tzortzakis, Detection of Harmful Residues in Honey Using Terahertz Time-Domain Spectroscopy, Applied Spectroscopy, Vol. 67(11), 2013, pp. 1264–1269.

[102]   L. Jiusheng, Optical Parameters of Vegetable Oil Studied by Terahertz Time-Domain Spectroscopy, Applied Spectroscopy, Vol. 64(2), 2010, pp. 231–234.

[103]   Y. Wang, Q. Wang, Z. Zhao, A. Liu, Y. Tian, and J. Qin, Rapid Qualitative and Quantitative Analysis of Chlortetracycline Hydrochloride and Tetracycline Hydrochloride in Environmental Samples Based on Terahertz Frequency-Domain Spectroscopy, Talanta, Vol. 190, 2018, pp. 284–291.

[104]   C. Jördens, and M. Koch, Detection of Foreign Bodies in Chocolate with Pulsed Terahertz Spectroscopy, Optical Engineering, Vol. 47(3), 2008, 037003.

Part III

Advances in the physical layer and network layer of nano-EM communication

Chapter 6

Channel modelling for electromagnetic nano-communication

*Rui Zhang[1], Ke Yang[2], Qammer H. Abbasi[3]
and Akram Alomainy[4]*

In vivo wireless nano-sensor networks (iWNSNs) have been presented as a mode to provide fast and accurate disease diagnosis and treatment. It will have promising applications in the biomedical field, such as intra-body health monitoring, drug delivery systems, immune system support mechanisms and artificial bio-hybrid implants. Despite the fact that nano-technology has advanced dramatically, enabling communication among nano-machines is still a major challenge. With respect to the electromagnetic (EM) communication approach, EM communication inside the body is frequency sensitive and depends on the nature of the medium they pass through; therefore, the operating frequency has an important effect on the communication channel. The miniaturisation of a conventional metallic antenna to meet the size requirement of a nano-sensor results in very high resonant frequencies, in the order of several hundreds of THz. At such frequencies, metals do not behave as perfect electric conductors, but exhibit a complex conductivity. This enables the propagation of confined EM modes at the surface of the antenna, which are commonly referred to as surface plasmon polariton (SPP) waves. Inspired by this phenomenon, a novel plasmonic nano-antenna for wireless communication among nano-devices has been recently proposed. The developments of miniature plasmonic signal sources, antennas and detectors, are expected to enable the wireless communications among *in vivo* nano-devices at the THz band [1,2].

Considering *in vivo* nano-communication, the THz band is the prior candidate because it is safe, due to its non-ionisation characteristics for biological tissues, and it is less susceptible to some of the propagation effects (i.e., Rayleigh scattering) [3,4]. Some studies have been conducted on the applicability of the infrared and optical transmission in a certain range of frequency windows (400–750 THz) [5].

[1]Department of Information and Electronic, Beijing Institute of Technology, Beijing, China
[2]School of Marine Science and Technology, Northwestern Polytechnical University, Xi'an, China
[3]James Watt School of Engineering, University of Glasgow, Glasgow, UK
[4]School of Electronic Engineering and Computer Science, Queen Mary University of London, London, UK

Intra-body wireless communication at the optical frequencies has been studied in [6] showing that the propagation of EM waves at optical frequencies inside the human body is mainly affected by scattering. In order to deeply exploit and further promote the potential of iWNSNs at the THz band for practical applications, accurate assessment of the propagation of THz wave in the biological medium is highly demanded. Accurate channel models are necessary to optimise the system parameters and build reliable, efficient and high-performance communication systems. Particularly, developing and assessing such models are imperative for achieving high data rates, targeting link budgets, determining optimal operating frequencies and designing efficient antenna and transceivers.

6.1   End-to-end terahertz propagation

6.1.1   *Molecular absorption*

Molecular absorption is a process in which the EM energy is partially transformed into kinetic energy and passes internally to vibrating molecules [7], which can be described by the absorption coefficient. Because the vibration frequencies at which a given molecule resonates change with the internal structure of the molecule, this quantity depends on the frequency. Given the absorption coefficient, the amount of incident EM radiation that is capable of propagating through the absorbing medium at a given frequency can be calculated. This parameter is defined by transmittance, which is obtained by using the Beer-Lambert law [8] as

$$\tau(r,f) = e^{-\alpha(f)r} \tag{6.1}$$

where f is the frequency of the EM wave, r is the total path length and $\alpha(f)$ stands for the absorption coefficient.

Molecular absorption causes attenuation to signals and can be obtained from the transmittance of the medium τ as [9]:

$$PL_{abs}(r,f) = \frac{1}{\tau(r,f)} = e^{\alpha(f)r} \tag{6.2}$$

In this study, the communication medium for iWNSNs is focused on human blood, skin and fat tissues, and their absorption coefficients at the frequency band of interest are shown in Figure 6.1. These are preliminary systematic measurements of human tissue optical properties in the THz band. The details on the measurement and signal processing can be found in [10,11]. The study has some limitations in maintaining the hydration level of the tested tissues, and the absorption coefficient presented may include attenuation due to scattering as well as absorption, which generate some variations along the frequency. Compared to the absorption coefficient of water vapour provided in [9], on the one hand, the absorption coefficient in human tissues can be thousands of times than that in the air at the same frequency. On the other hand, different from the thousand resonant peaks of water vapour over

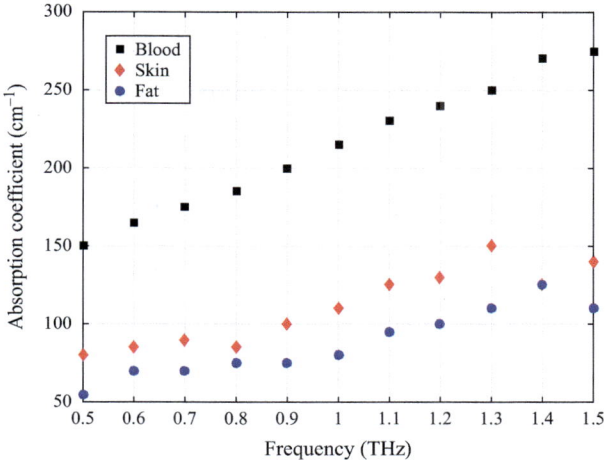

Figure 6.1   Molecular absorption coefficient as a function of the frequency of human blood, skin and fat (reproduced from [10,11])

the THz band, the absorption coefficient of human tissues increases with a frequency much more steadily. It gives *in vivo* THz communication some peculiar behaviours, which will be presented in the following sections.

6.1.2   Scattering

In addition to the molecular absorption, EM waves suffer from the scattering due to the deflection of the beam caused by the microscopic non-uniformities present in the human body. Scattering depends on the diameters of the particles and the wavelength of the THz wave. The aforementioned two parameters aid in determining the kind of scattering that the communication will suffer from. Depending on the diameter of the particles and the wavelength of the THz wave, there are three kinds of scattering as follows [12]:

- Rayleigh scattering (particles diameter < wavelength of THz wave)
- Mie scattering (particles diameter ≈ wavelength of THz wave)
- Specular and geometric scattering (particles diameter > wavelength of THz wave)

The models for scattering by particles and cells in the human body have been analysed in [13]. It is found that the scattering coefficients for blood, skin and water are much less than the absorption coefficient at the same frequency in the THz band. Thus, compared to the molecular absorption, the scattering effect can be neglected, and the absorption is the dominant attenuation for the *in vivo* EM nano-communication at the THz frequencies.

6.1.3   Path loss

The path loss in human tissues can be divided into two parts: the spreading loss and molecular absorption loss [14]. The spreading loss is a part of attenuation caused by the expansion of a wave propagating through the medium, and it can be calculated from the modified Friis transmission equation [8]:

$$PL_{spr}(r,f) = \left(\frac{4\pi nfr}{c}\right)^2 \qquad (6.3)$$

where n is the refractive index of the medium in the THz frequencies, $4r^2$ denotes the isotropic expansion term and $4\pi(nf/c)^2$ stands for the frequency-dependent receiver antenna aperture term. Considering the attenuation, the total path loss can be described as

$$PL(r,f) = PL_{spr}(r,f)PL_{abs}(r,f) = \left(\frac{4\pi nfr}{c}\right)^2 e^{a(f)r} \qquad (6.4)$$

and the expected received signal power can be represented by [15]:

$$P_R(r) = \int_B S(f)\left(\frac{c}{4\pi nfr}\right)^2 e^{-a(f)r} df \qquad (6.5)$$

where $S(f)$ is the transmitted signal power spectral density (PSD) from the transmitter antenna and B is the channel bandwidth. Here, the antennas of the transmitter and receiver are assumed to be ideal isotropic.

The path loss was validated by simulating the plane wave propagating in the tissue using CST Microwave Studio in [16]. The 3D schematics of the simplified model of the propagation of the plane wave in human tissues is shown in Figure 6.2. Several probes have been equally spaced in order to monitor the variation of the E-field with the distance from the source. A perfect matched layer (PML) has been considered as the boundary condition in the direction of propagation and the periodic conditions in other directions. A plane wave has been defined as the propagating wave travelling in the direction of $+z$ while the E-field

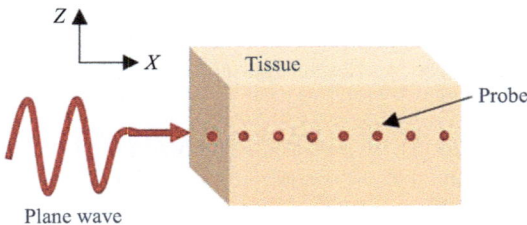

Figure 6.2   3D schematics of the simplified model of the propagation of the plane wave in human tissues

is pointing to $+x$. The obtained numerical absorption loss results match well with the analytical results, as shown in [16].

The dependency of the channel path loss on the distance and the frequency is illustrated in Figure 6.3. Due to the significantly high absorption coefficient in human tissues, the path loss suffers more from the molecular absorption loss than the spreading loss. More specifically, at the same frequency and transmission distance, the exponential loss caused by the molecular absorption is more than double

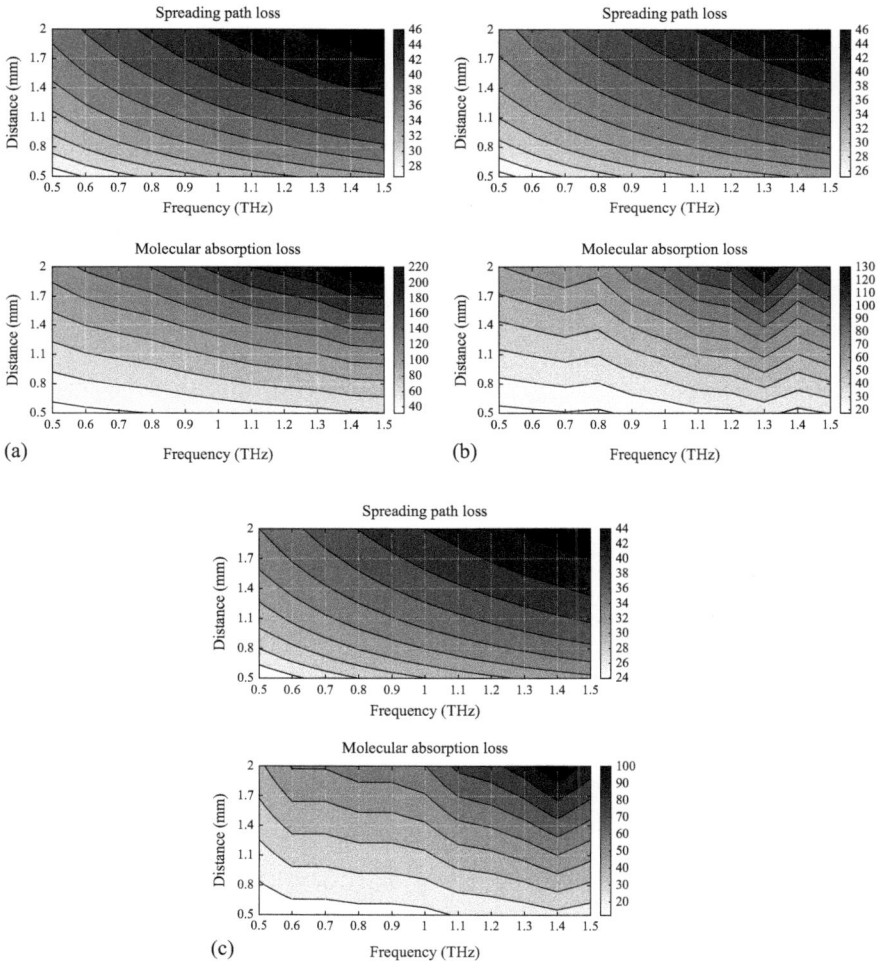

Figure 6.3 Spreading path loss and molecular absorption loss as a function of frequency and propagation distance in different tissues: (a) blood; (b) skin and (c) fat

the spreading loss contributions to the path loss. Furthermore, it can be seen that the path loss increases with the increase in both the transmission distance and the frequency.

6.1.4 Molecular absorption noise

The noise in the THz band is primarily contributed by the molecular absorption noise. This kind of noise is caused by vibrating molecules which partially re-radiate the energy that has been previously absorbed [9]. It is developed that the molecular absorption noise is contributed by the background noise and the self-induced noise as [17]:

$$N(r,f) = N_b(f) + N_s(r,f) \qquad (6.6)$$

The background noise caused by the radiation of the medium can be described by Planck's function [18]:

$$B(T_0,f) = \frac{2h\pi f^3}{c^2}(e^{hf/k_B T_0} - 1)^{-1} \qquad (6.7)$$

where k_B is the Boltzmann constant, h is the Planck constant, T_0 is the effective temperature of the atmosphere and c is the speed of light in vacuum. Planck's function is multiplied with π to transform the unit from W/Hz/cm^2/sr to W/Hz/cm^2.

For simplicity, the transmission medium is assumed to be an isothermal and homogeneous layer with a thickness d. As mentioned earlier, this background noise is generated by the radiation of the local sources of the medium, and it is assumed that this radiation is only from the original energy state of the molecules before transmission happens; thus, it is independent on the transmitted signal. The background noise can be described as [18]:

$$N_b(f) = \int_0^d B(T_0,f)\alpha(f)e^{-\alpha(f)s}ds = B(T_0,f)(1 - e^{-\alpha(f)d}) \simeq B(T_0,f) \qquad (6.8)$$

The integral in (6.8) describes the noise intensity at the centre of a sphere with a radius d, given that all the points s in the medium contribute to the noise intensity. Since it is obtained by Planck's function, the unit of the background noise PSD is W/Hz/cm^2. The background noise can be further approximated by taking into account the (ideal) antenna aperture term $c^2/(4\pi f_0)^2$ to get the background noise PSD with the unit W/Hz [19]. Furthermore, with regard to the *in vivo* scenario, the speed of light in the human body could change with the composition of the medium and the frequency of the THz wave. Thus, the background noise PSD can be written as

$$N_b(f) = B(T_0,f)\frac{c^2}{4\pi n_0^2 f_0^2} \qquad (6.9)$$

where f_0 is the design centre frequency and n_0 is the corresponding refractive index of the medium.

In terms of the induction mechanism of the self-induced noise, the internal vibration of the molecules results in the emission of EM radiation at the same frequency of the incident waves that provoked this motion [7,9]. It is obtained with the assumption that all the absorbed energy from the transmitted signal received at the receiver would turn into self-induced noise as [17]:

$$N_s(r,f) = S(f)(1 - e^{-\alpha(f)r})\left(\frac{c}{4\pi nrf}\right)^2 \tag{6.10}$$

In addition to the molecular absorption noise, there are other noise sources that can affect communication performance, such as the device noise. A number of prototypes of antenna, emitting at the THz band, available today are built with conventional materials. In this case, the Johnson–Nyquist thermal noise should be taken into account as a source of noise [20]. However, with the development of new materials, it is possible that the thermal noise can be neglected with the use of graphene and its derivatives to create THz antennas. Graphene-based nano-structures allow the ballistic transport of electrons, leading to a very low thermal noise in the nano-antennas and nano-transceivers; thus, it is reasonably expected that the thermal noise is minor [21,22]. Therefore, the molecular absorption noise, in reaction in response to the absorption of EM waves by molecules in the channel, is the dominant contributor to the noise at the receiver.

It can be seen from (6.9) that the background noise depends on the temperature and composition of the medium. It is assumed that the human tissues are isothermal; thus, the background noise changes slightly with the refractive index of different transmission media.

For the general communication systems, besides the effect of both path loss and noise, communication capabilities are also strictly influenced by the distribution of power transmission in the frequency domain $S(f)$ [23]. Here, two communication schemes (namely, flat and pulse-based) are considered to analytically investigate the communication performance of the *in vivo* EM nano-communication at the THz band.

6.1.4.1 Flat power distribution

In the simplest case, the total transmitted signal power P_T is uniformly distributed over the entire operative band (e.g., 0.5–1.5 THz). Thus, the corresponding transmitted signal PSD is

$$S_{flat}(f) = P_T/B \text{ for } f \in B, \quad 0 \text{ otherwise} \tag{6.11}$$

6.1.4.2 Pulse-based power distribution

The transmitted signal can be modelled with an nth derivative of a Gaussian-shape: $\phi(f) = (2\pi f)^2 n e^{(-2\pi\sigma f)^2}$ [3]. Thus, the signal PSD can be expressed as [9]:

$$S_p^{(n)}(f) = a_0^2 \phi(f) \tag{6.12}$$

where σ and a_0^2 are the standard deviation of Gaussian pulse and a normalising constant, respectively. Considering that $\int_{f_m}^{f_M} S_P^{(n)}(f)df = P_T$, the normalising constant is obtained as [3]:

$$a_0^2 = \frac{P_T}{\int_{f_m}^{f_M} \phi(f)} \tag{6.13}$$

To keep the numerical results realistic, and in light of the state of the art in nano-transceivers, in this study, signal power is adopted with the total energy equal to 0.1 aJ and the pulse duration be 100 fs. For the Gaussian pulse-based transmission scheme, the derivative order n and the standard deviation of the Gaussian pulse σ are set to 6 and 0.15, respectively. The background noise PSD, self-induced noise PSD and molecular absorption noise PSD for human blood, skin and fat tissues are shown in Figures 6.4, 6.5 and 6.6, respectively.

From Figure 6.4, it can be seen clearly that for both flat and Gaussian pulse-based power distribution schemes, the background noise PSD is almost the same as it is in different kinds of tissue, because the slight difference of refractive index does not play a significant role in (6.9). In addition, the background noise is independent on the path length.

Figure 6.5 shows that the self-induced noise PSD has a steady change with frequency, which is different from the abrupt fluctuation of THz communication in the air as shown in [9]. The reason is that the molecular absorption coefficient steadily increases over the frequency of interest. Furthermore, the self-induced noise PSD is inversely proportional to the path length. Considering the power allocation, the self-induced noise of THz communication in the same kind of tissue using flat communication decreases with the increase of the frequency, while the one in the Gaussian pulse-based scenario increases with the rise of the frequency. It is mainly because of the transmitted signal changes in diverse trends with frequency.

It is shown in Figure 6.6 that the molecular absorption noise PSD using flat communication is higher than the Gaussian pulse-based scheme in the same tissue type, with the same transmitted signal power and at the same communication distance and frequency. In terms of the variations among different tissues, when sharing the same transmitted signal power using Gaussian pulse communication, the molecular absorption noise in blood is less than that in skin and fat, because the absorption coefficient and refractive index increase with the water concentration in the medium and blood has higher water proportion than skin and fat, comparatively. However, in flat communication scenario, this relation depends on the optical parameters of the specific tissue and the transmitted signal power.

6.1.5   Signal-to-noise ratio

In order to analyse the communication performance, especially the achievable communication range, SNR of the *in vivo* communication channel is investigated, which can be written as a function of the transmission distance and frequency:

$$SNR(r,f) = \frac{S(f)PL^{-1}(r,f)}{N(r,f)} \tag{6.14}$$

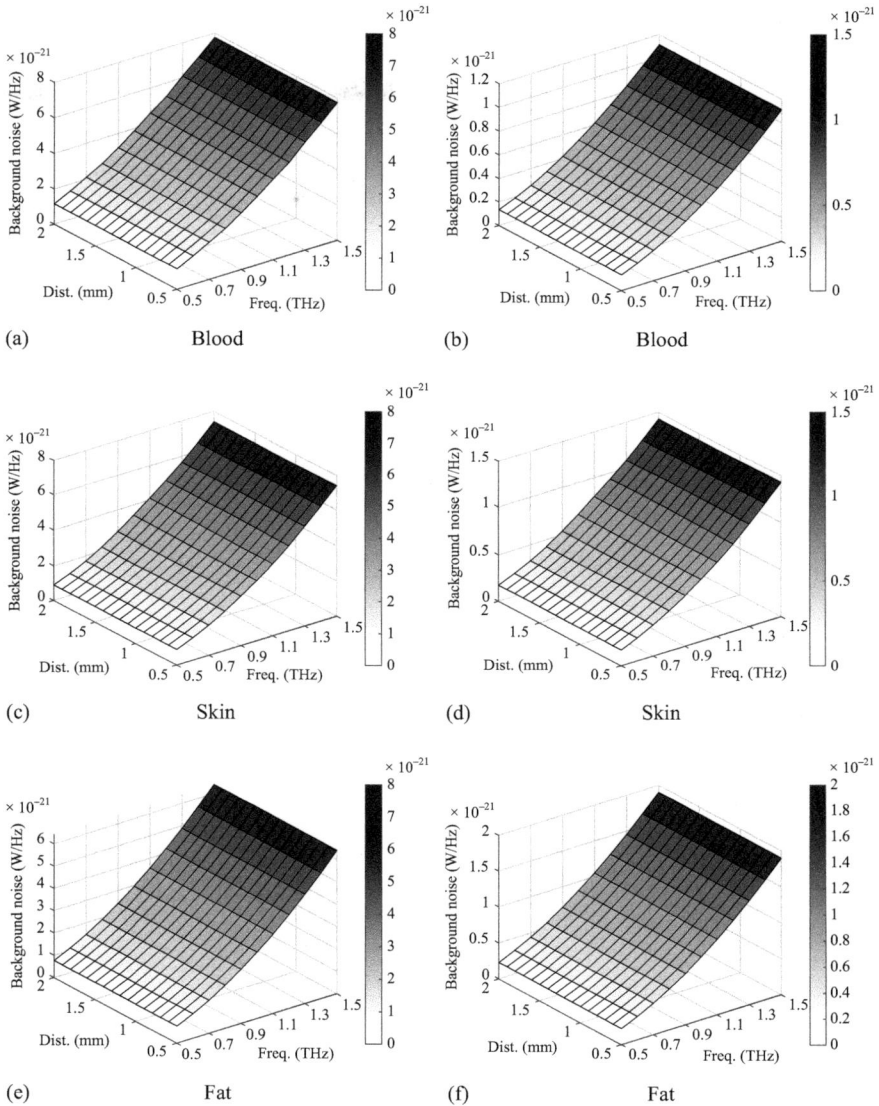

(a)   Blood   (b)   Blood

(c)   Skin   (d)   Skin

(e)   Fat   (f)   Fat

Figure 6.4   *Background noise PSD for THz communication inside human blood, skin and fat tissues. The left column is for flat power distribution, while the right column is for Gaussian pulse-based power distribution*

where $S(f)$ stands for the PSD of the transmitted signal, $PL(r,f)$ denotes the channel path loss and $N(r,f)$ refers to the molecular absorption noise PSD.

For the considered human tissue types, i.e., blood, skin and fat, the two communication schemes (flat and Gaussian-shaped pulses) result in different SNR values as shown in Figure 6.7. It can be seen that SNR is inversely proportional to

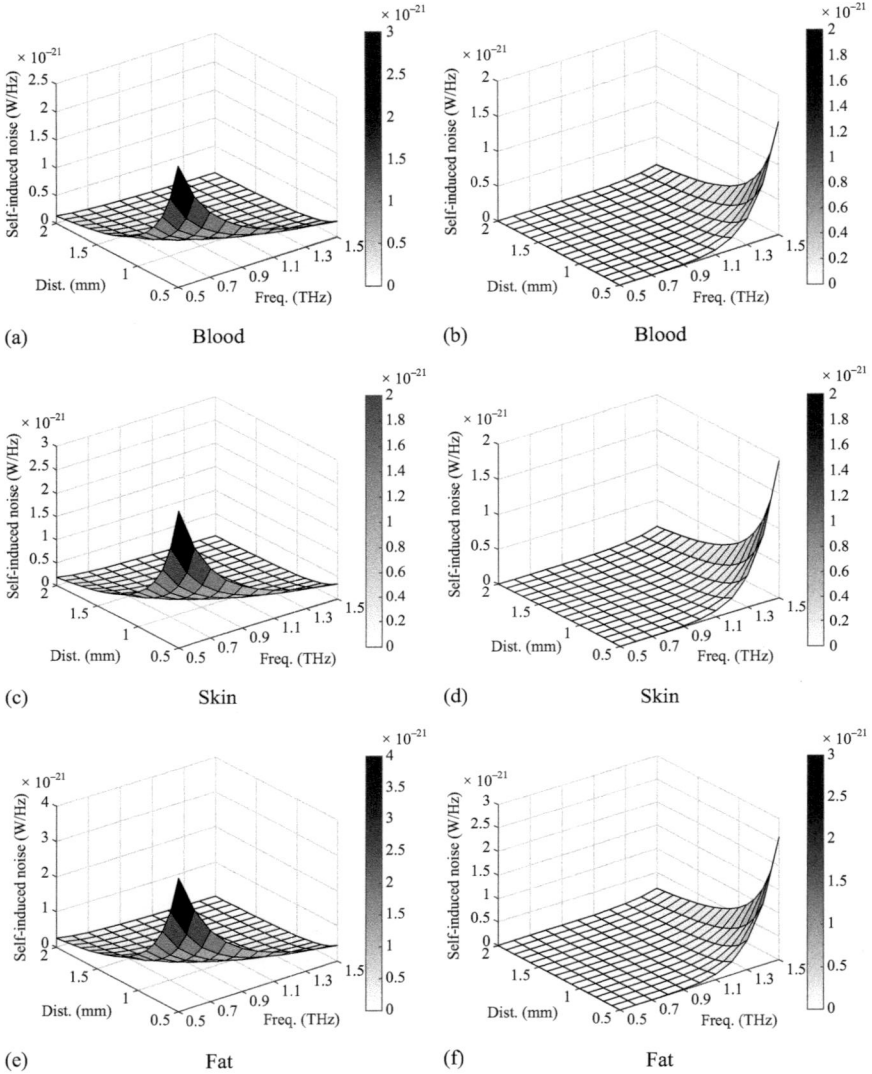

*Figure 6.5   Self-induced noise PSD for THz communication inside human blood,
skin and fat tissues. The left column is for flat power distribution,
while the right column is for Gaussian pulse-based power distribution*

the distance regardless of the composition of the human tissues and the transmitted
signal. Besides, the THz communication in blood has the lowest SNR, and that in
skin has the second-lowest while that in fat has the highest value. It is indicated that
SNR degrades rapidly with the increase of the water concentration in the communication medium.

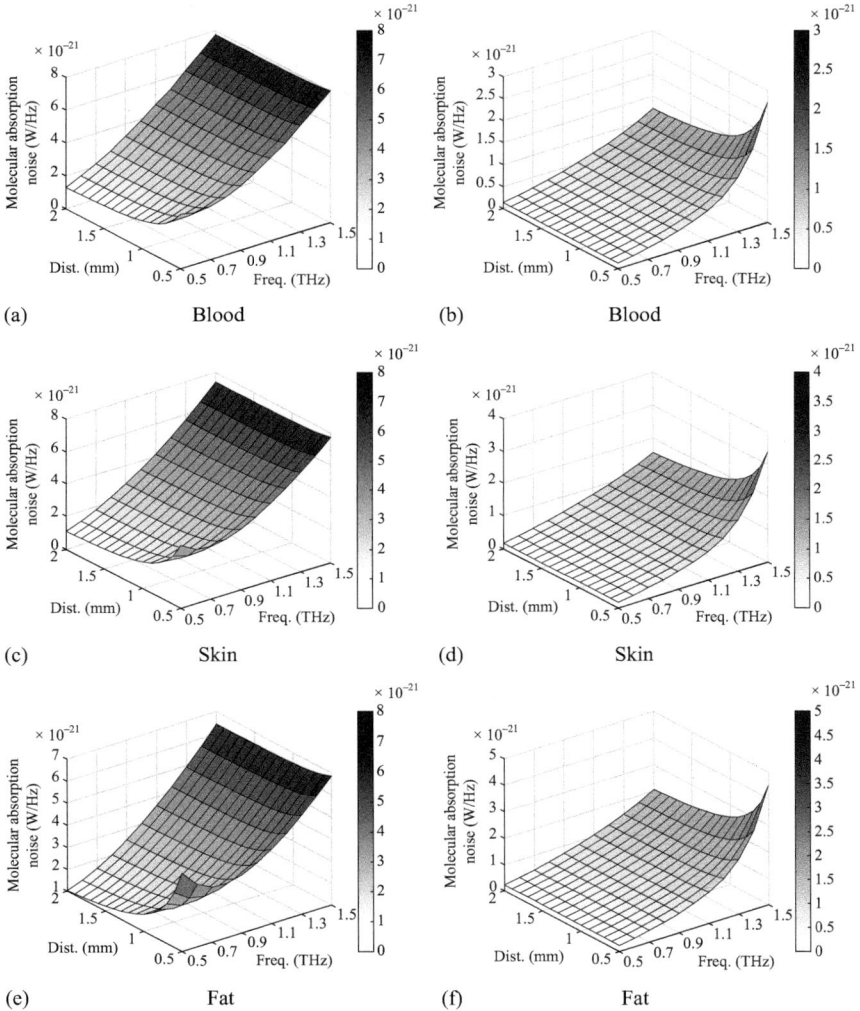

Figure 6.6   Molecular absorption noise PSD for THz communication inside human blood, skin and fat tissues. The left column is for flat power distribution, while the right column is for Gaussian pulse-based power distribution

With respect to the flat power distribution, SNR inversely decreases with the THz signal frequency in all three tissues. Specifically, take blood as an example, SNR of the THz communication at 0.5 THz at the path length of 0.5 mm is −35.4 dB, and it decreases to −57.5 and −76.6 dB when the frequency becomes 1 and 1.5 THz, respectively. Moreover, this difference is expanding with the increase of the transmission distance. SNR of the communication inside blood with a

(a)   Blood   (b)   Skin

(c)   Fat

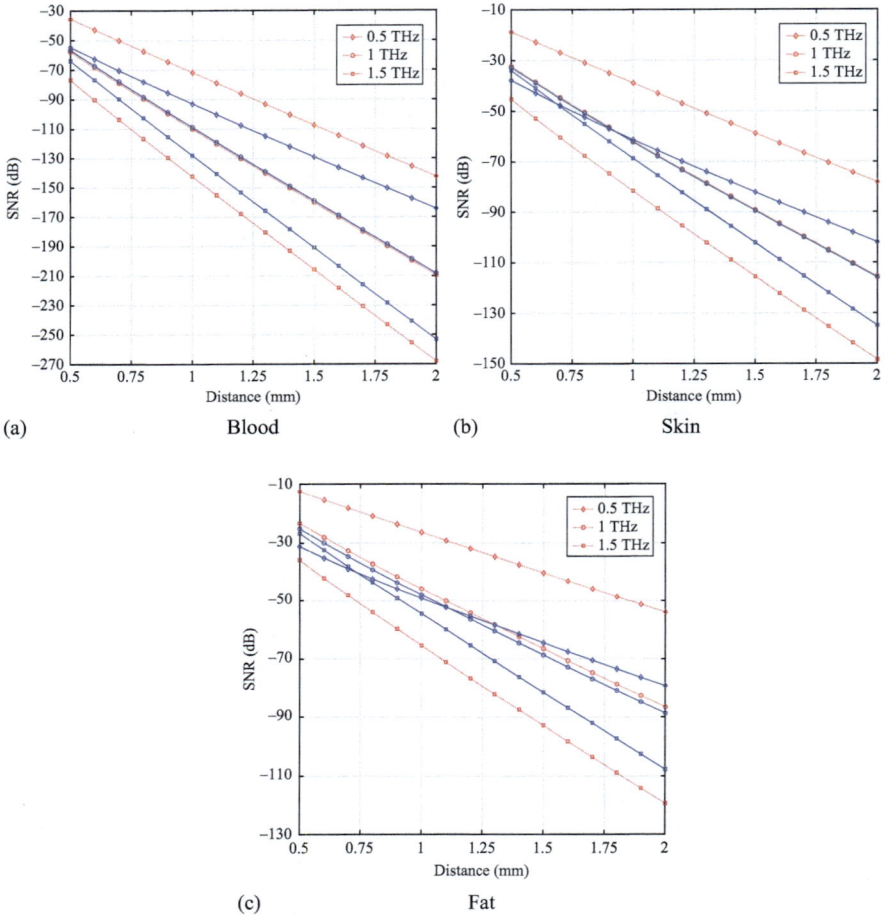

Figure 6.7   SNR as a function of the transmission distance for different power allocation schemes in human blood, skin and fat tissues. In the figures, the dotted line denotes flat power distribution, while the solid line refers to Gaussian pulse-based power distribution

distance of 2 mm at 0.5 THz is −142.3 dB, while it is −209.4 and −267.9 dB at 1 and 1.5 THz, respectively. When it comes to Gaussian pulse-based power distribution, there is no consistent relation between SNR and the frequency, and SNR highly depends on the specific frequency, path length and tissue composition. For instance, when the communication in the skin is less than 1 mm, the transmission of 1 THz has the highest SNR as compared to 1.5 and 0.5 THz. However, SNR of communication with 0.5 THz wave can be higher than the ones of 1 and 1.5 THz when the path length extends beyond 1 mm.

In light of the state of the art in communication devices and in an effort to make the iWNSNs realistic, it can be concluded that the maximum achievable transmission distance of *in vivo* nano-communication at the THz band can be approximately 1–2 mm. More specific transmission distance depends on the composition of the transmission medium, especially the water concentration of the medium; the operation band of *in vivo* nano-communication can be limited to the lower band of the THz band (lower than 1 THz). Although the communication distance is strongly limited in the THz band, this distance is estimated to be sufficient for the dense iWNSNs. For example, in iWNSNs, the density of nano-nodes is extremely high, which is in the range of hundreds of nano-sensors per square millimetre [24], hence making the communication distance acceptable in iWNSNs.

6.1.6 Time spread on–off keying

Due to the fact that nano-devices in WNSNs are highly energy constrained with limited capabilities, it is technologically challenging for a nano-transceiver to generate a high-power carrier frequency in the THz band. Thus, the best modulation option for WNSNs is carrier-less pulse-based modulation [25]. In light of the state of the art in graphene-based nano-electronics, a transmission scheme for nano-devices, based on the transmission of 100-fs long pulses by following an on–off keying modulation spread in time is proposed [26], named TS-OOK. A logical '1' is transmitted by using a fs-long pulse and a logical '0' is transmitted as silence, i.e., the nano-device remains silent when a logical zero is transmitted. These very short pulses can be generated and detected with nano-transceivers based on graphene and high-electron-mobility materials such as gallium nitride or indium phosphide [27]. To the authors' best knowledge, currently, TS-OOK is the most promising communication scheme for resource-constrained nanonetworks.

Utilising TS-OOK, only the background noise is present, when transmitting silence for a logical '0', and the noise power can be obtained as

$$N_0 = \int_B B(T_0, f) \frac{c^2}{4\pi(n_0 f_0)^2} df \tag{6.15}$$

where B is the channel bandwidth.

When transmitting a pulse for a logical '1', both the background noise and the self-induced noise are present, which can be written as

$$N_1 = \int_B \left(B(T_0, f) \frac{c^2}{4\pi(n_0 f_0)^2} + S(f)(1 - e^{-a(f)r}) \left(\frac{c}{4\pi r n f} \right)^2 \right) df \tag{6.16}$$

6.1.7 Information rate

TS-OOK is taken into consideration to evaluate the effect of the molecular absorption noise on the performance of THz communication inside the human body. The maximal mutual information rate is calculated by the transmitted and received signals to quantify the potential of THz band for communication inside the human body. The maximum achievable information rate in bit/symbol of a communication system for a specific modulation scheme is given by the well-known Shannon limit theorem [28]:

$$IR = \max_X\{H(X) - H(X \mid Y)\} \text{ [bit/symbol]} \tag{6.17}$$

where X refers to the source of information, Y stands for the output of the channel, $H(X)$ refers to the entropy of the source X and $H(X \mid Y)$ stands for the conditional entropy of X given Y or the equivocation of the channel. When using TS-OOK, the source X can be modelled as a discrete binary random variable. Therefore, the entropy of the source $H(X)$ is given by [28]:

$$H(X) = -\sum_{m=0}^{1} p_X(x_m)\log_2 p_X(x_m) \tag{6.18}$$

where $p_X(x_m)$ refers to the probability of transmitting the symbol $m = 0, 1$, i.e., the probability to stay silent or to transmit a pulse, respectively.

Since the molecular absorption noise can be modelled as additive coloured Gaussian noise (ACGN) [29], the probability density function (PDF) of the molecular absorption noise at the receiver conditioned to the transmission of the symbol x_m is given by [30]:

$$f_N(n \mid X = x_m) = \frac{1}{\sqrt{2\pi N_m}}e^{-n^2/2N_m} \tag{6.19}$$

where n refers to noise and N_m refers to the molecular absorption noise power when symbol m (0/1) is transmitted in (6.15) and (6.16).

When considering a 1-bit hard receiver based on power detection, the system becomes a binary asymmetric channel (BAC) and Y is a discrete random variable. This channel can be fully characterised by the four transition probabilities [30]:

$$\left.\begin{aligned}
p_Y(Y = 0 \mid X = 0) &= \int_{th_1}^{th_2} f_Y(y \mid X = 0)dy \\
p_Y(Y = 1 \mid X = 0) &= 1 - p_Y(Y = 0 \mid X = 0) \\
p_Y(Y = 0 \mid X = 1) &= \int_{th_1}^{th_2} f_Y(y \mid X = 0)dy \\
p_Y(Y = 1 \mid X = 1) &= 1 - p_Y(Y = 0 \mid X = 1)
\end{aligned}\right\} \tag{6.20}$$

where th_1 and th_2 are two threshold values and $f_Y(y \mid X = x)$ is the PDF of the channel output Y conditioned to the transmission of the symbol $X = x$, which is given by [30]:

$$f_Y(y \mid X = x_m) = \delta(y - a_m) * f_N(n = y \mid X = x_m) = \frac{1}{\sqrt{2\pi N_m}}e^{-(n-a_m)^2/2N_m} \tag{6.21}$$

where δ stands for the Dirac delta function and a_m refers to the received symbol amplitude, obtained from (6.5).

Contrary to the classical symmetric additive Gaussian noise channel, in the asymmetric channel, there are two points at which $f_Y(y \mid X = 0)$ and $f_Y(y \mid X = 1)$ intersect. It is considered that these thresholds be defined for the case without interference [30]. Therefore, th_1 and th_2 can be analytically computed from the intersection between two Gaussian distribution $N(0, N_0)$ and $N(0, N_1)$, respectively, which results in [30]:

$$
th_{1,2} = \frac{a_1 N_0}{N_0 - N_1}
$$
$$
\pm \frac{\sqrt{2 N_0 N_1^2 \log \frac{N_1}{N_0} - 2 N_0^2 N_1 \log \frac{N_1}{N_0} + a_1^2 N_0 N_1}}{N_0 - N_1} \tag{6.22}
$$

The equivocation of the channel $H(X \mid Y)$ for the BAC is given by [30]:

$$
H_{BAC}(X \mid Y) = \sum_{y=0}^{1} \sum_{x=0}^{1} p_Y(Y = y \mid X = x) p_X(X = x)
$$
$$
\log_2 \left(\frac{\sum_{q=0}^{1} p_Y(Y = y \mid X = q) p_X(X = q)}{p_Y(Y = y \mid X = x) p_X(X = x)} \right) \tag{6.23}
$$

Finally, the maximum achievable information rate in bit/second is obtained by multiplying the rate in bit/symbol (6.17) by the rate at which symbols are transmitted, $R = 1/T_s = 1/(\beta T_p)$, where T_s is the time between symbols, T_p is the pulse length and β is the ratio between them. Assuming that the $BT_p \simeq 1$, where B stands for the channel bandwidth, the rate in bit/second is given by [30]:

$$
IR_u = \frac{B}{\beta} IR_{u_{sym}} \tag{6.24}
$$

If $\beta = 1$, i.e., all the symbols (pulses or silences) are transmitted in a burst, and the maximum rate per nano-device is achieved, provided that the incoming information rate and the read-out rate to and from the nano-transceiver can match the channel rate. By increasing β, the single-user rate is reduced, but the requirements on the transceiver are greatly relaxed.

Here, the bandwidth of the THz communication channel is considered to be 1 THz, and the information rate for three considered human tissues has been studied and the results are shown in Figure 6.8. It can be seen clearly that the information rate decreases with the rise of the transmission distance in both flat and pulse-based scheme. Specifically, with respect to flat distribution, the information rate for THz communication inside human fat is about 10 Gigabit per second (Gbps) at 0.5 mm communication distance and it decreases to about 70 Megabit per second (Mbps) when the distance further increases to 2 mm; the information rate in skin drops from 2 Gbps to 10 Mbps, while it goes down from 0.5 Gpbs to 2 Mbps in blood. Differently, the information rate for THz communication with Gaussian pulse has a slight difference in various tissues and it decreases from around 6 Gbps at a path length of 0.5 mm to 30 Mbps at 2 mm. The main reason for such high

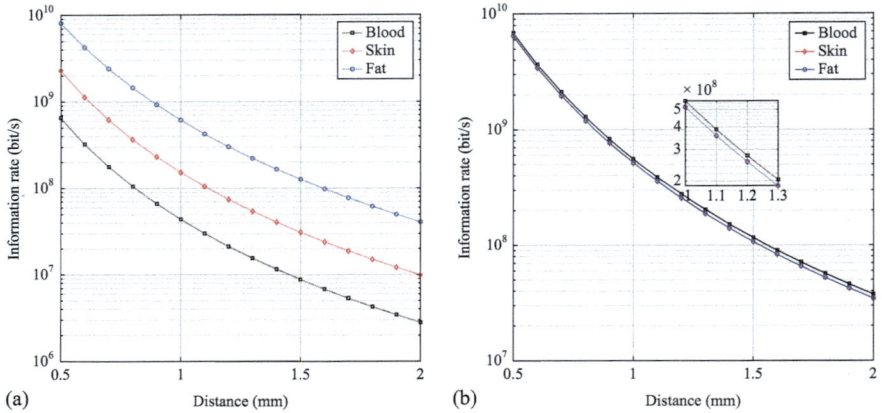

Figure 6.8   Information rate as a function of transmission distance in different human tissues with (a) flat and (b) Gaussian pulse-based power distribution scheme

information rate is the extremely high bandwidth of THz communication. The obtained information rate indicates that complex tasks can be completed in the envisioned nano-communication inside the human body by the high ability of successfully transmitting information over the communication channel.

6.2   Multi-user terahertz propagation

The end-to-end model is an ideal simplification of the communication channel, whereas in reality, the communication environment is much more complicated and the requirement of nanonetworks widely change across applications. For example, in WNSNs, very high node densities, in the order of hundreds of nano-sensors per square millimetre, are needed to overcome the limited sensing range of individual devices [24]. Moreover, different types of nano-devices could be interleaved and cooperated to conduct more complicated tasks, resulting in up to thousands of nano-machines per square millimetre. With respect to dense iWNSNs, multi-user interference will occur, when symbols from different transmitters reach the receiver at the same time and overlap. Therefore, not only the path loss and molecular absorption noise, but also the multi-user interference could be the major impairment that degrades the performance of *in vivo* nano-communication at the THz band. In addition to the interference caused by different nodes in the communication area, with the utilisation of TS-OOK, collisions between symbols can also occur. These collisions result in interference, which imposes a strict limitation on the communication channel parameters and requirements, especially the achievable distance at which nano-machines can communicate.

In order to evaluate the potential of the THz communication inside the human body, it is significantly important to conduct network-level analysis and characterise the system performance while using TS-OOK as a communication scheme. While assessing the link performance of the wireless communication, the probability distribution and mean values of signal-to-interference-plus-noise ratio (SINR) are quantified as fundamental metrics.

6.2.1 Interference model

Generally, a random deployment of nodes in R^2 as shown in Figure 6.9 is used for performance assessment of cellular, ad hoc and device-to-device networks [31]. The targeted receiver is assumed to be at the centre of the disc, while the targeted transmitter locates at a distance r_0 from the receiver. All the other nodes in this field are considered as interfering nodes for the targeted receiver. Following most studies, the Poisson point process (PPP) is utilised to provide the first-order approximation of nodes' positions within a disc of radius R [32]. Thus, the probability of finding M nodes in the area $A(R)$ can be represented as [33]:

$$P[M \mid A(R)] = \frac{(\lambda \pi R^2)^M}{M!} e^{-\lambda \pi R^2} \tag{6.25}$$

where λ refers to the node density in nodes/m^2. The mean and variance of the number of an interferer M can be written as [33]:

$$E[M] = \lambda \pi R^2, \quad \sigma^2[M] = \lambda \pi R^2 \tag{6.26}$$

The SINR is an important metric to evaluate the system performance of the dense nanonetworks. The instantaneous frequency-dependent SINR is defined as

$$\mathrm{SINR}(\vec{r}, S, f, \lambda) = \frac{P_R(r_0, S, f)}{I(\vec{r}, S, f, \lambda) + N(\vec{r}, S, f, \lambda)} \tag{6.27}$$

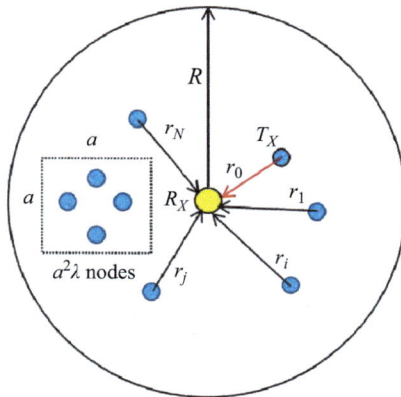

Figure 6.9 Nano-devices deployment inside the body for multi-user scenario

where $\vec{r}$ is the vector of distances $r_i, i = 1, 2, \dots, M$, standing for the separation distances between the interferer and the targeted receiver, S is the PSD of the transmitted signal power and M is the number of interfering nodes. f is the operating frequency and λ is the intensity of interferer. $P_R(r_0, S, f)$ refers to the targeted receiver signal power, while $I(\vec{r}, S, f, \lambda)$ denotes the aggregate power of the interfering signals at the targeted receiver and $N(\vec{r}, S, f, \lambda)$ is the noise power at the targeted receiver including all the noise power caused by the targeted transmitter and interferer. In this chapter, it is assumed that there are no power control capabilities, which means that $S_i = S_j = S, i, j = 0, 1, \dots, M$ [34,35]. For simplicity, in the following, we drop arguments of notation that are often silently assumed, f, S and λ. The aggregate interference from M sources is given by [34]:

$$I(\vec{r}) = \sum_{i=1}^{M} \int_B S(f) \left(\frac{c}{4\pi n f}\right)^2 r_i^{-2} e^{-\alpha(f) r_i} df \tag{6.28}$$

where $S(f)$ is the transmitted signal PSD from the transmitter antenna. The noise power is written as a summation of the noise caused by both the targeted transmitter and the interfering nodes in the communication area [36]:

$$N(\vec{r}) = N_m + \sum_{i=1}^{M} \int_B S(f) \left(\frac{c}{4\pi n f}\right)^2 r_i^{-2} (1 - e^{-\alpha(f) r_i}) df \tag{6.29}$$

where N_m is the molecular absorption noise power obtained from (6.16). Substituting (6.29) and (6.28) into (6.27) gives

$$\mathrm{SINR}(\vec{r}) = \frac{P_R}{N_m + \sum_{i=1}^{M} \int_B S(f) \left(\frac{c}{4\pi n f}\right)^2 r_i^{-2} df} \tag{6.30}$$

whereas in (6.30) P_R is a constant value that can be estimated for any given distance r_0. The second term in the denominator of (6.30) is the only random term, which is given as

$$X(\vec{r}) = \sum_{i=1}^{M} A r_i^{-2}, \quad A = \int_B S(f) \left(\frac{c}{4\pi n f}\right)^2 df \tag{6.31}$$

where X is caused by the presence of the interferer in the communication medium, which includes both the received signal power generated by the interferer nodes and the noise power caused by the interferer signal.

Another important observation is that the distances from any interferer to the targeted receiver are independent and identically distributed (IID). For a sufficiently large number of users, the central limit theorem can be invoked and the Gaussian assumption can be made for X, when estimating the aggregate interference [26]. Therefore, the moments of the interference from a single node are first determined and then the central limit theorem is applied to approximate the aggregated interference.

With respect to the dense nanonetworks with the adoption of TS-OOK as the communication scheme, a collision between symbols will occur when they reach the receiver at the same time and overlap. The probability of having an arrival during T_s seconds is a uniform random probability distribution with PDF equal to $1/T_s$ [26]. Therefore, for a specific transmission, a collision will happen with a probability $2T_p/T_s$ (with an assumption that a correlation-based energy detector is used at the receiver) [26].

It is noted that not all types of symbols harmfully collide, only pulses (logical '1's) create interference because the molecular absorption noise is signal power dependent. It is assumed that all nano-nodes in the transmission area share the same pulse transmitting probability. Therefore, the node density parameter λ in (6.26) can be replaced by [30]:

$$\lambda' = \lambda_T \left(2T_p/T_s\right)p_1 \tag{6.32}$$

where λ_T refers to the density of active nodes in nodes/m^2, T_p is the symbol duration, T_s is the time between symbols and p_1 denotes the probability of a nano-machine to transmit a pulse. This expression highlights the fact that transmission of logical '0' does not generate interference to other ongoing transmissions. Both the interference caused by the interfering nodes in the transmission area and the interference generated by the utilisation of TS-OOK are taken into account in (6.32).

6.2.1.1 Single node interference model

For a PPP, any given number of interferers in a disc of radius R is independently and uniformly distributed. Therefore, the distance to the targeted receiver, denoted as a random variable D, has the same PDF for any interferer node:

$$f_D(r) = 2r/R^2, \ 0 < r < R \tag{6.33}$$

Now, consider a random variable $G = 1/D^2$. Under the Poisson assumption, the moments of G can be written as [37]:

$$E\left[G^\theta\right] = \int_0^R \frac{2x}{R^2} \left(\frac{1}{x^2}\right)^\theta dx \tag{6.34}$$

The integral does not converge because it is unbounded approaching zero from the right. To deal with this issue, it is assumed that the transmitters cannot be located closer than a certain very small distance a from the receiver. This assumption is warranted from the practical point of view, especially, taking into account that a can be chosen as small as required [35]. Thus, the distribution of the distance to the targeted receiver could be approximated as the distance from a point arbitrarily distributed in the region bounded by two concentric circles of radius a and R, $R > a$, to their common centre. It is known to be [34]:

$$f_D(r) = 2r/\left(R^2 - a^2\right), \ a < r < R \tag{6.35}$$

The first moment of variable G is computed as

$$E[G] = \int_a^R \frac{2x}{(R^2 - a^2)} \frac{1}{x^2} dx = \frac{2(\ln R - \ln a)}{R^2 - a^2} \tag{6.36}$$

Similarly, the variance of G is calculated to be

$$\sigma^2[G] = \frac{1}{a^2 R^2} - 4 \left(\frac{\ln R - \ln a}{R^2 - a^2} \right)^2 \tag{6.37}$$

The interference model for a single node has been obtained, and then it is moved to the aggregate interference.

6.2.1.2   Aggregate interference model

It is assumed that the number of interferers is exactly k, which results in the conditional moment of X to be [34]:

$$E[X(\vec{r}) \mid M = k] = A \sum_{i=1}^k E[G_i] = AkE[G] \tag{6.38}$$

Denoting $P_r(M = k) = p_k$ and moment of X can be calculated as

$$E[X(\vec{r})] = A \sum_{k=0}^{\infty} p_k k E[G] = AE[G]E[M] \tag{6.39}$$

Similarly, the second conditional moment of X is given by

$$E[X^2(\vec{r}) \mid M = k] = A^2 E\left[\left(\sum_{i=1}^k r_i^{-2} \right)^2 \right]$$
$$= A^2 \sum_{i=1}^k \sum_{j=1}^k (E[G_i]E[G_j] + K_{ij}) \tag{6.40}$$

where $K_{ij} = Cov(G_i, G_j)$ is the pairwise covariance. G_i and G_j are pairwise independent $K_{ij} = 0$, $i = 1, 2, \ldots, k, ij$ and $K_{ii} = \sigma^2[G_i], i = 1, 2, \ldots, k$ [30]. Further, since all G_i identically distributed:

$$E[G_i] = E[G_j], \quad \sigma^2[G_i] = \sigma^2[G_j] \tag{6.41}$$

Thus, the second moment of X can be calculated as

$$E[X^2(\vec{r})] = A^2((E[G])^2 E[M^2] + \sigma^2[G]E[M]) \tag{6.42}$$

Then, the variance of X can be found as

$$\sigma^2[X(\vec{r})] = A^2((E[G])^2 \sigma^2[M] + \sigma^2[G]E[M]) \tag{6.43}$$

In (6.31) the moments of random variable X are being calculated; thus, PDF of SINR in dB can be obtained based on conventional methods of finding distributions of functions of random variables. Recall that PDF of a random Y, $w(f)$, expressed as monotonous function $y = \phi(x)$ of another random variable X with PDF $f(x)$ is given by [38]:

$$w(y) = f(\psi(y))|\psi'(y)| \tag{6.44}$$

where $x = \psi(y) = \phi^{-1}(x)$ is the inverse function.

The inverse of $y = \phi(x) = 10\log_{10}(P_R/(N_m + x))$ is unique and monotonous and given by $x = \psi(y) = P_R 10^{-y/10} - N_m$. The modulo of the derivative is $|\psi'(y)| = |P_R 10^{-y/10} \ln(10^{-1/10})|$. Substituting these into (6.44), PDF of the logarithm of SINR is

$$w_{\log S}(y) = \frac{|P_R 10^{-y/10} \ln\left(10^{-1/10}\right)|}{\sqrt{2\pi}\sigma} e^{-\left(P_R 10^{-y/10} - N_m - \mu\right)^2/2\sigma^2} \tag{6.45}$$

where $\mu = E[X(\vec{r})], \sigma^2 = \sigma^2[X(\vec{r})]$ can be obtained in (6.39) and (6.43).

6.2.2 SINR distribution

Based on the aforementioned derived models, it is noted that the distribution and average values of SINR are directly dependent on the node density and probability of transmitting pulses. In this section, the effect of these parameters on the system performance of iWNSNs inside three human tissues (blood, skin and fat) is analytically investigated. The simulation environment of the following analytical study is summarised in Table 6.1.

6.2.2.1 Effect of the node density

With respect to TS-OOK, in order to satisfy the requirement that the time between symbols should be much longer than the symbol duration, it is envisioned that $T_s/T_p = 100$. Aforementioned, the node density of iWNSNs can be hundreds of

Table 6.1 Simulation environment

Parameters	Definition
$R = 3$ mm	The radius of the considered disc
$r_0 = 1$ mm	The distance between the targeted transmitter and the targeted receiver
$T_0 = 310$ K	The temperature of human tissues, and tissues are assumed as isothermal
$P_T = 0.1$ aJ	The total energy of the transmitted Gaussian pulse
$B = 1$ THz	The bandwidth of the transmitted Gaussian pulse
$T_p = 100$ fs	The pulse duration of the transmitted Gaussian pulse
$n = 6$	The derivative order of the transmitted Gaussian pulse
$\sigma = 0.15$	The standard deviation of the transmitted Gaussian pulse

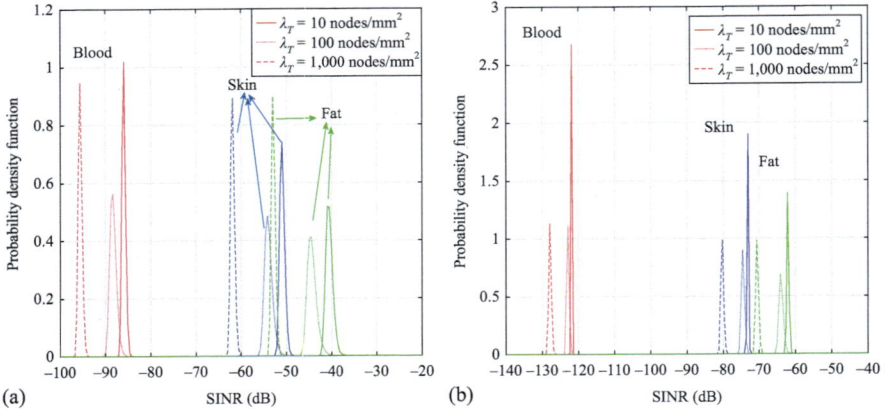

Figure 6.10   *The probability density function of SINR in human blood, skin and fat tissues for different node densities λ_T when $p_1 = 0.5$ with (a) flat power distribution and (b) Gaussian pulse-based power distribution*

nano-sensors per millimetre square. Therefore, $\lambda_T = 10, 100$ and $1,000$ nodes/mm^2 are chosen to evaluate the effect of the interfere density on the system performance.

Figure 6.10 shows the results for SINR distribution at different node densities with a specific signal transmission probability $p_1 = 0.5$ for THz wave communicating inside human blood, skin and fat tissues at 1 mm.

As expected, it can be concluded that for a specific probability of transmitting pulses, the SINR decreases significantly with the increase in node density for both power allocation scheme. Specifically, SINR created by a Poisson field of nano-devices with parameter $\lambda_T = 10$ nodes/mm^2, which are operating under the previous conditions in human blood, has an average power of approximately -85.9 dB. When the node density increases to $\lambda_T = 100$ nodes/mm^2 and $\lambda_T = 1,000$ nodes/mm^2, the values reduce to -88.3 and -95.6 dB, respectively. The reason is that collisions occur with a higher probability when interfere nodes grow in the communication area. Regarding Gaussian pulse-based communication, the average SINR drops from -128 to -123 dB and then to -122 dB when λ_T increases from 10 to 100 and then to $1,000$ nodes/mm^2. SINR of the THz communication in human skin and fat tissues experiences a similar trend.

6.2.2.2   Effect of the probability of transmitting pulses

Figure 6.11 illustrates the effect of the probability of transmitting pulses on the distribution of SINR in the communication system. In terms of the flat communication, for a specific node density $\lambda_T = 100$ nodes/mm^2, by increasing p_1 from 0.1 to 0.5 and then to 0.9, the average SINR of nanonetworks in human blood decreases from -88 to -90 dB and then to -91.2 dB. With regard to Gaussian pulse-based communication, the average SINR in blood degrades about 1 dB from -121 to -122 dB when p_1 increases from 0.1 to 0.5 and drops to -123 dB when

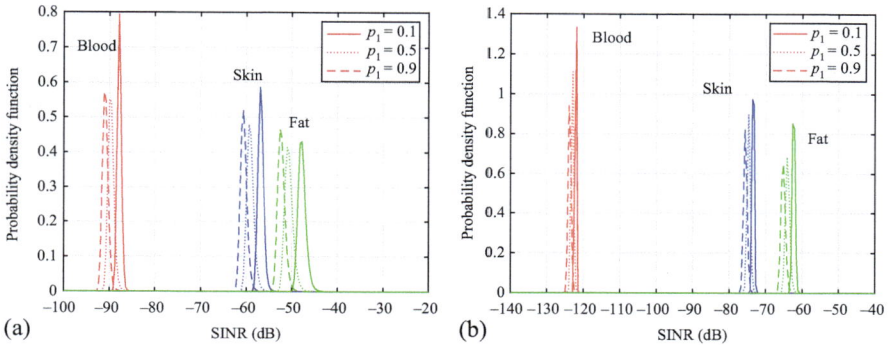

Figure 6.11   *The probability density function of SINR for different probabilities of transmitting pulses p_1 when $\lambda_T = 100$ nodes/mm^2 in human blood, skin and fat with (a) flat communication and (b) Gaussian pulse communication*

$p_1 = 0.9$. These results emphasise the fact that the molecular absorption noise and multi-user interference are directly dependent on the transmitted signal; thus, the transmitting pulses potentially degrade the communication system performance.

The obtained results imply that by controlling the probability of the transmission of the pulse and node density, the interference can be reduced and SINR of the *in vivo* nano-communication system can be bettered. Practically, a trade-off can be made between node density and the probability of the transmission of the pulse to achieve the expected communication performance.

6.3   Summary

In this chapter, the path loss and molecular absorption noise models for the *in vivo* THz communication are introduced. Moreover, the analytical results on SNR and information rate with flat and Gaussian pulse-based power allocation scheme are presented. It indicates that the maximum achievable transmission distance of *in vivo* THz communication should be restrained to approximately 1–2 mm, and more specific transmission distance limitation depends on the composition of the transmission medium, especially the water concentration of the medium. The operation band of iWNSNs is limited to the frequencies lower than 1 THz. The information rate decreases steadily with the increase in the transmission distance regardless of the type of the medium and can reach several Gbps when the transmission distance is 0.5 mm. Afterwards, an interference model for iWNSNs with the utilisation of TS-OOK is developed based on the mathematical apparatus of stochastic geometry. The performance of the multi-user communication inside human blood, skin and fat is comparatively illustrated, showing that blood is the worst performing scenario because of higher water concentration than skin and fat. In all three kinds of tissues, the obtained results show that high node density and pulse transmission probability would potentially decrease SINR of the system and impair the system performance. Flat and Gaussian-pulse based power distribution scheme behaves differently in different tissues in the THz frequencies. Therefore, a proper power allocation should be selected based on the specific application. The presented results provide an important basis for more practical network-level modelling, stimulating further research on simple, reliable and energy efficient communication protocols and coding schemes.

References

[1]   Akyildiz IF, and Jornet JM. Electromagnetic wireless nanosensor networks. Nano Communication Networks. 2010;1(1):3–19.

[2]   Tamagnone M, Gomez-Diaz J, Mosig JR, *et al*. Reconfigurable terahertz plasmonic antenna concept using a graphene stack. Applied Physics Letters. 2012;101(21):214102.

[3]   Piro G, Yang K, Boggia G, *et al*. Terahertz communications in human tissues at the nanoscale for healthcare applications. IEEE Transactions on Nanotechnology. 2015;14(3):404–406.

[4]   Abbasi QH, El Sallabi H, Chopra N, *et al*. Terahertz channel characterization inside the human skin for nano-scale body-centric networks. IEEE Transactions on Terahertz Science and Technology. 2016;6(3):427–434.

[5]   Nafari M, and Jornet JM. Metallic plasmonic nano-antenna for wireless optical communication in intra-body nanonetworks. In: Proceedings of the 10th EAI International Conference on Body Area Networks. ICST (Institute for Computer Sciences, Social-Informatics and Telecommunications Engineering); 2015. pp. 287–293.

[6]   Elayan H, Shubair RM, Jornet JM, *et al*. Terahertz channel model and link budget analysis for intrabody nanoscale communication. IEEE Transactions on Nanobioscience. 2017;16(6):491–503.

[7]   Jornet JM, and Akyildiz IF. Channel capacity of electromagnetic nanonetworks in the terahertz band. In: 2010 IEEE International Conference on Communications (ICC). IEEE. pp. 1–6.

[8]   Fox M. Optical properties of solids. Oxford master series in condensed matter physics. Oxford University Press, Oxford; 2001.

[9]   Jornet JM, and Akyildiz IF. Channel modeling and capacity analysis for electromagnetic wireless nanonetworks in the terahertz band. IEEE Transactions on Wireless Communications. 2011;10(10):3211–3221.

[10]   Fitzgerald A, Berry E, Zinov'ev N, *et al*. Catalogue of human tissue optical properties at terahertz frequencies. Journal of Biological Physics. 2003; 29(2–3):123–128.

[11]   Berry E, Fitzgerald AJ, Zinov'ev NN, *et al*. Optical properties of tissue measured using terahertz-pulsed imaging. In: Medical Imaging 2003: Physics of Medical Imaging, vol. 5030. International Society for Optics and Photonics; 2003. pp. 459–471.

[12] Bohren CF, and Huffman DR. Absorption and scattering of light by small particles. New York: John Wiley & Sons; 2008.

[13] Elayan H, Shubair RM, and Jornet JM. Bio-electromagnetic THz propagation modeling for in vivo wireless nanosensor networks. In: 2017 11th European Conference on Antennas and Propagation (EUCAP). IEEE; 2017. pp. 426–430.

[14] Yang K, Pellegrini A, Munoz MO, Brizzi A, Alomainy A, and Hao Y. Numerical analysis and characterization of THz propagation channel for body-centric nano-communications. IEEE Transactions on Terahertz Science and Technology. 2015;5(3):419–426.

[15] Kokkoniemi J, Lehtomäki J, Umebayashi K, *et al.* Frequency and time domain channel models for nanonetworks in terahertz band. IEEE Transactions on Antennas and Propagation. 2015;63(2):678–691.

[16] Yang K. Characterisation of the in-vivo terahertz communication channel within the human body tissues for future nano-communication networks. Queen Mary University of London; 2016.

[17] Zhang R, Yang K, Abbasi QH, *et al.* Analytical modelling of the effect of noise on the terahertz in-vivo communication channel for body-centric nano-networks. Nano Communication Networks. 2017;15:59–68.

[18] Chandrasekhar S. Radiative transfer. New York: Courier Corporation; 2013.

[19] Kokkoniemi J, Lehtomäki J, and Juntti M. A discussion on molecular absorption noise in the terahertz band. Nano Communication Networks. 2016;8:35–45.

[20] Boronin P, Moltchanov D, and Koucheryavy Y. A molecular noise model for THz channels. In: 2015 IEEE International Conference on Communications (ICC). IEEE; 2015. pp. 1286–1291.

[21] Geim AK, and Novoselov KS. The rise of graphene. Nature Materials. 2007; 6(3):183–191.

[22] Pal AN, and Ghosh A. Ultralow noise field-effect transistor from multilayer graphene. Applied Physics Letters. 2009;95(8):082105.

[23] Goldsmith A. Wireless communications. Cambridge: Cambridge University Press; 2005.

[24] Jornet JM, and Akyildiz IF. Femtosecond-long pulse-based modulation for terahertz band communication in nanonetworks. IEEE Transactions on Communications. 2014;62(5):1742–1754.

[25] Zarepour E, Hassan M, Chou CT, *et al.* Performance analysis of carrier-less modulation schemes for wireless nanosensor networks. In: 2015 IEEE 15th International Conference on Nanotechnology (IEEE-NANO). IEEE; 2015. pp. 45–50.

[26] Jornet JM, and Akyildiz IF. Low-weight channel coding for interference mitigation in electromagnetic nanonetworks in the terahertz band. In: 2011 IEEE International Conference on Communications (ICC). IEEE; 2011. pp. 1–6.

[27] Jornet JM, and Akyildiz IF. Graphene-based plasmonic nano-transceiver for terahertz band communication. In: The 8th European Conference on Antennas and Propagation (EuCAP 2014). IEEE; 2014. pp. 492–496.

[28]   Shannon CE. A mathematical theory of communication. ACM SIGMOBILE Mobile Computing and Communications Review. 2001;5(1):3–55.

[29]   Jornet JM, and Akyildiz IF. Information capacity of pulse-based wireless nanosensor networks. In: 2011 8th Annual IEEE Communications Society Conference on Sensor, Mesh and Ad Hoc Communications and Networks (SECON). IEEE; 2011. pp. 80–88.

[30]   Jornet JM. Low-weight error-prevention codes for electromagnetic nano-networks in the terahertz band. Nano Communication Networks. 2014; 5(1):35–44.

[31]   ElSawy H, Hossain E, and Haenggi M. Stochastic geometry for modeling, analysis, and design of multi-tier and cognitive cellular wireless networks: A survey. IEEE Communications Surveys & Tutorials. 2013;15(3):996–1019.

[32]   Andrews JG, Ganti RK, Haenggi M, *et al.* A primer on spatial modeling and analysis in wireless networks. IEEE Communications Magazine. 2010; 48(11):156–163.

[33]   Cox DR, and Isham V. Point processes. vol. 12. Boca Raton, FL: CRC Press; 1980.

[34]   Petrov V, Moltchanov D, and Koucheryavy Y. Interference and SINR in dense terahertz networks. In: 2015 IEEE 82nd Vehicular Technology Conference (VTC Fall). IEEE; 2015. pp. 1–5.

[35]   Petrov V, Moltchanov D, and Koucheryavy Y. On the efficiency of spatial channel reuse in ultra-dense THz networks. In: 2015 IEEE Global Communications Conference (GLOBECOM). IEEE; 2015. pp. 1–7.

[36]   Zhang R, Yang K, Abbasi Q, *et al.* Analytical characterisation of the terahertz in-vivo nano-network in the presence of interference based on TS-OOK communication scheme. IEEE Access. 2017;5:10172–10181.

[37]   Papoulis A, and Pillai SU. Probability, random variables, and stochastic processes. New York: McGraw-Hill Education; 2002.

[38]   Feller W. An introduction to probability theory and its applications: volume I. vol. 3. John Wiley & Sons, London; 1968.

Chapter 7

Modulation, coding, and synchronization techniques for nano-electromagnetic communications in terahertz band

Muhammad Mahboob Ur Rahman[1], Qammer Hussain Abbasi[2], Akram Alomainy[3], Hasan Tahir Abbas[2] and Muhammad Ali Imran[2]

7.1 Introduction

This chapter aims to provide insights into the modulation and coding schemes (MCS) tailored for—and synchronization issues faced by—the nano-scale electromagnetic communication systems operating in the terahertz (THz) band. The MCS is of paramount importance for the works dealing with rate and power adaptation by the transmitter when channel state information at transmitter (CSIT) of the link is available. Synchronization, on the other hand, refers to frequency, phase, and timing synchronization and is the pre-requisite for many modern signal processing operations, advanced multi-antenna techniques, and advanced modulation techniques. Some prominent examples that require synchronization include decoding at the receiver, multi-input, multi-output (MIMO) systems, beamforming, and orthogonal frequency division multiplexing (OFDM).

Ever since its inception, nano-scale communication in the THz band has attracted limited attention (mainly due to the grave challenges in fabricating the nano-scale transceivers). Therefore, only a handful of works on modulation schemes and MCS tailored for—and synchronization challenges faced by—the nano-scale communication in the THz band have been reported in the literature so far. However, due to the recent breakthrough in the fabrication of novel (graphene-based) 2D materials, the interest in nano-scale communication in the THz band has increased by many folds. To this end, in addition to the research contributions by

[1]Electrical Engineering Department, Information Technology University, Lahore, Pakistan
[2]James Watt School of Engineering, University of Glasgow, Glasgow, UK
[3]School of Electronic Engineering and Computer Science, Queen Mary University of London, London, UK

the researchers from academia, the industry has also come forward to express their interest in this novel communication regime by preparing a standard, namely, "IEEE standard for high data rate wireless multi-media networks" (a.k.a. the IEEE 802.15.3 standard). The standard was approved and published online in 2016, and later, revised in 2017. This chapter, therefore, reviews the works dealing with the MCS and synchronization needs of the communication systems operating in the THz band in a comprehensive manner. More importantly, this chapter does an extensive (but not exhaustive) study of the literature on MCS and synchronization techniques for the wireless links operating at microwave frequencies. With this, the authors aim to provide their opinion regarding the portability of the existing MCS and synchronization techniques to THz frequencies and discuss the potential implementation challenges if any.

7.2   Modulation in the THz band

Due to extremely large bandwidths (up to 100 GHz) offered by the THz band and due to potential limitations on the computational complexity of nano-scale transceivers, modulation schemes in the THz band come in two different flavors: (i) pulse-based modulation and (ii) traditional carrier-based modulation (CBM). The IEEE 802.15.3 standard accordingly defines two different PHY (i.e., physical) layers for the nano-scale communication systems operating in the THz band: (i) a THz physical layer with single-carrier or THz-SC, and (ii) a THz physical layer with on–off keying or THz-OOK [1]. In other words, the THz-SC PHY layer represents the CBM, while the THz-OOK PHY layer represents the pulse-based modulation.

7.2.1   Pulse-based modulation

The ultra-high bandwidth (much greater than the bandwidth of ultra-wideband systems) enables pulse-based modulation (PBM) whereby very short-lived pulses (of duration 1 fs or less) are transmitted by the nano-transmitter to implement, say, using an OOK scheme [2]. The receiver's task is then to do energy detection (which could be threshold-based or matched filter-based) for decoding purposes [3]. The PBM is used to ease the transmitter and receiver design whenever they have stringent limits on their computational complexity. Note that the IEEE standard 802.15.3 utilizes the OOK modulation scheme in its THz-OOK PHY layer to realize the pulse-based modulation.

7.2.2   Carrier-based modulation

For the traditional CBM in the THz band, a standard set of modulation schemes could be used, e.g., phase shift keying (PSK), quadrature amplitude modulation (QAM), OFDM, frequency shift keying, amplitude shift keying (ASK), pulse amplitude modulation, etc. To be concrete, the IEEE 802.15.3 standard suggests utilizing the following modulation schemes: binary PSK (BPSK), quadrature PSK (QPSK), 8-PSK, 8-APSK, 16-QAM, and 64-QAM.

Note that the ultra-high bandwidth of the nano-scale communication systems operating in the THz band implies ultra-small time slots (on the order of nano-seconds or so) which puts a hard limit on electronics design whereby the current state-of-the-art field-programmable gate arrays (FPGAs) and analogue-to-digital converters (ADCs) still fall short of meeting such small timelines. One potential solution, therefore, could be to investigate mixed (analogue/digital) architectures for precoding and decoding, in line with the literature on communication systems utilizing the millimeter (mm) wave band [4].

Furthermore, the device form-factor constraints might limit the power budget at the nano-transmitter and hence the signal-to-noise ratio (SNR) at the nano-receiver (despite the fact that the length of the nano-scale communication channel is a few centimeters maximum). This, in turn, limits the decoding complexity at the nano-receiver as the receiver might not be able to implement, say, the Viterbi algorithm with a long Trellis graph, to do maximum likelihood sequence estimation (MLSE) for 100 symbols. Thus, for M-ary modulation schemes in the THz band, M might be chosen to be a small number (compared to communication at microwave frequencies where $M = 512$ is commonplace in the latest Wi-Fi and cellular standards).

In the literature on nano-scale communication systems operating in the THz band, computational complexity is not the only metric to realize different modulation paradigms (i.e., PBM vs. CBM). That is, modulation schemes have also been designed keeping various other objectives in mind, e.g., energy-efficiency [5], interference [6], and transmission of multiple symbols via a single pulse in a PBM system [7].

When the nano-transmitter is intelligent, i.e., it is capable of switching between PBM and CBM from one slot to another, one important design objective for the intelligent nano-receiver is to discriminate between PBM and CBM on the fly. To this end, the work in [8] provides a systematic mechanism which allows the nano-receiver to decide between PBM or CBM via simple energy testing (which is based on the fact that for a given observation interval, the energy of the CBM signal is significantly greater than the energy of the PBM signal). Additionally, when the mode of modulation is CBM, Iqbal *et al.* [8] perform modulation classification (to find the modulation scheme used by the nano-transmitter during the current slot) by constructing a Gaussian mixture model + expectation-maximization + Kullback–Leibler divergence-based framework.

Recently, there has been a growing interest to design optical transceivers and monolithic microwave integrated circuits (MMICs) working at THz frequencies that utilize a menu of modulation schemes and are capable of providing data rates on the order of few tens to hundreds of Gb/s [9–30]. To this end, a comprehensive summary of the various modulation schemes and the corresponding data rates reported by the works [9–30] is (culled from [7]) provided in Table 7.1.

To sum things up, the general consensus of the community is to use CBM for the long-range/macro-scale systems (where nano-transmitter and nano-receiver are

*Table 7.1   Various modulation schemes and the corresponding achievable data
rates reported by the communication systems operating in the THz band
(culled from [7])*

Frequency (THz)	Data rate (Gb/s)	Distance (m)	Modulation	Reference
0.125	10	200	ASK	[9]
0.25	8	0.5	ASK	[10]
0.2	1	2.6	ASK	[11]
0.12	10	5,800	ASK	[12]
0.3	0.096	0.7	64-QAM	[13]
0.625	2.5	<10	Duobinary	[14]
0.22	15–40	10	OOK	[15]
0.24	25	60	OOK	[16]
0.0875	100	1.2	16-QAM	[17]
0.135	10	0.2	ASK	[18]
0.3	24	0.5	ASK	[19]
0.146	1	0.025	OOK	[20]
0.22	30	20	ASK	[21]
0.542	2	0.01	ASK	[22]
0.14	10	1,500	16-QAM	[23]
0.24	30	40	8-PSK	[24]
0.196	0.1	0.5	QPSK	[25]
0.34	3	0.3	16-QAM	[26]
0.3	24	0.3	ASK	[27]
0.3	48	1	OOK	[28]
0.237	100	20	16-QAM	[29]
0.4	40	2	ASK	[30]

separated by a few meters or so), while pulse-based modulation schemes are used
for short-range/micro-scale systems (where nano-transmitter and nano-receiver are
separated by a few centimeters or so).

7.3   Channel coding schemes in the THz band

It is anticipated that the nano-receivers could only offer limited digital signal processing capabilities; therefore, it is reasonable to assume that only those channel
codes with relatively short (finite) block-length will find their way into nano-scale
communication systems operating in THz. The literature on channel coding
schemes which are tailored for the THz band is extremely scarce. Specifically, the
authors are only aware of a single work [31] that utilizes a low-weight coding
scheme to mitigate the interference seen by a THz link. As for the IEEE
802.15.3 standard, low-density parity-check (LDPC) codes and Reed–Soloman
(RS) codes with the following three configurations are recommended: 14/15-rate
LDPC (1440,1344), 11/14-rate LDPC (1440,1056), and 11/14-rate RS (240,224)-
code.

Since MCS are almost always studied together for power and rate adaptation when CSIT is available, we discuss MCS schemes next.

7.4 MCS in the THz band

As mentioned at the beginning of the chapter, modulation and coding schemes are of paramount importance for rate and power adaptation by the transmitter when CSIT of the link is available. The underlying principle of all the MCS schemes is to construct a lookup table which helps the transmitter to select a particular rate and channel coding scheme based upon the quantized CSIT [32].

Again, only a handful of works have been reported that discuss the MCS tailored for the THz band. Specifically, Moshir and Singh [7] proposed an MCS-based rate-adaptation technique for a time-domain THz system. Another closely related work is [33] that does rate maximization via modulation selection and power allocation when CSIT is available. Last but not least, Figure 7.1 (culled from the IEEE 802.15.3 standard document) shows a lookup table suggested by the IEEE 802.15.3 for implementation purposes. One could see from Figure 7.1 that CSIT is quantized into a 4-bit number—the so-called MCS identifier—whose value is then used to select a modulation scheme as well as a rate for the forward error correction (FEC) (i.e., channel coding) scheme. The theoretically achievable data rates for a given MCS identifier value, for different bandwidths, with and without the pilot word (PW) are also listed. It is straightforward to note that a higher SNR implies a larger value for the MCS identifier, and thus, the nano-transmitter could choose an M-ary modulation scheme with bigger M and an (n, k) channel code with a bigger k/n ratio.

7.5 Synchronization in the THz band

It is anticipated that a multitude of applications utilizing the THz band will need synchronization as a pre-requisite, e.g., (distributed) ultra-massive MIMO [34], etc. Note that the synchronization, in its traditional sense, is applicable only to those THz systems that utilize CBM schemes. For the THz systems utilizing pulse-based modulation schemes, one needs to look at the problem of timing synchronization alone (as there are no carrier signals involved, thus, there are no frequency and phase offsets) so that both the nano-scale transmitter and nano-scale receiver have a consistent notion of the beginning of the time slots.

To date, only a handful of works have been reported in the literature that discuss synchronization needs of the THz systems. Gupta *et al.* [35] investigate the symbol time synchronization issues for a THz system that uses pulse-based modulation. To this end, they propose a scheme that iteratively estimates the symbol start time and reduces the observation window length for the symbol detector; the proposed scheme is analogue as it could be implemented with a combination of voltage-controlled delay lines and continuous-time moving-average symbol detectors. Although not a (physical layer) synchronization work, Xia *et al.* [36] present a

MCS identifier	Modulation	FEC rate	Bandwidth 2.16 GHz Data rate (Gb/s)		Bandwidth 4.32 GHz Data rate (Gb/s)		Bandwidth 8.64 GHz Data rate (Gb/s)		Bandwidth 12.96 GHz Data rate (Gb/s)		Bandwidth 17.28 GHz Data rate (Gb/s)		Bandwidth 25.92 GHz Data rate (Gb/s)		Bandwidth 51.84 GHz Data rate (Gb/s)		Bandwidth 69.12 GHz Data rate (Gb/s)	
			Without PW	With PW	Without PW	With PW	Without PW	With PW	Without PW	With PW	Without PW	With PW	Without PW	With PW	Without PW	With PW	Without PW	With PW
0	BPSK	11/15	1.29	1.13	2.58	2.26	5.16	4.52	7.74	6.78	10.33	9.04	15.49	13.55	30.98	27.11	41.30	36.14
1	BPSK	14/15	1.64	1.44	3.29	2.87	6.57	6.75	9.86	8.62	13.14	11.50	19.71	17.25	39.42	34.50	52.56	45.99
2	QPSK	11/15	2.58	2.26	5.16	4.52	10.33	9.03	15.49	13.55	20.65	18.07	30.98	27.10	61.95	54.21	82.60	72.28
3	QPSK	14/15	3.29	2.87	6.57	5.75	13.14	11.50	19.71	17.25	26.28	23.00	39.42	34.50	78.85	68.99	105.13	91.99
4	8-PSK	11/15	3.87	3.39	7.74	6.78	15.49	13.55	23.23	20.33	30.98	27.11	46.47	40.66	92.93	81.32	123.91	108.42
5	8-PSK	14/15	4.93	4.31	9.86	8.62	19.71	17.25	29.57	25.87	39.42	34.50	59.13	51.74	118.27	103.49	157.69	137.98
6	8-APSK	11/15	3.87	3.39	7.74	6.78	15.49	13.55	23.23	20.33	30.98	27.11	46.47	40.66	92.93	81.32	123.91	108.42
7	8-APSK	14/15	4.93	4.31	9.86	8.62	19.71	17.25	29.57	25.87	39.42	34.50	59.13	51.74	118.27	103.49	157.69	137.98
8	16-QAM	11/15	5.16	4.52	10.33	9.03	20.65	18.07	30.98	27.10	41.30	36.14	61.95	54.21	123.90	108.42	165.21	144.55
9	16-QAM	14/15	6.57	5.75	13.14	11.50	26.28	23.00	39.42	34.50	52.57	45.99	78.85	68.99	157.70	137.98	210.26	183.98
10	64-QAM	11/15	7.74	6.78	15.49	13.55	30.98	27.10	46.46	40.66	61.95	54.21	92.93	81.31	185.86	162.62	247.81	216.83
11	64-QAM	14/15	9.86	8.62	19.71	17.25	39.42	34.50	59.14	51.74	78.85	68.99	118.27	103.49	236.54	206.98	315.39	275.97

Figure 7.1   The MCS and data rates lookup table provided by the IEEE standard 802.15.3 (culled from [1])

medium access control (MAC) protocol which does link-layer synchronization to realize ultra-fast communication networks operating in the THz band. The authors claim that their proposed protocol could maximize the successful packet delivery probability without compromising the achievable throughput. On the radio frequency (RF) technology front, the design of Schottky-based mixers and oscillators that work efficiently till frequencies up to and beyond 2 THz is gaining attention from the researchers lately [37].

On the other hand, at microwave frequencies, the phenomena of synchronization are well-studied whereby the oscillator dynamics have been mathematically modeled and empirically verified [38–41].

Below, we describe the essential ingredients of the three (frequency, phase, and timing) synchronization problems when addressed at microwave frequencies, one by one.

7.5.1   Frequency synchronization

As hinted earlier, oscillator dynamics leads to randomly time-varying frequency offsets which in turn lead to several problems at the receiver, e.g., the rotation of the M-ary constellation at the receiver, inter-carrier interference in OFDM receivers, reduction in the received power/signal-to-noise ratio (SNR) gain at the receiver due to transmit beamforming, etc. To this end, modeling frequency drift as a random walk with Gaussian process noise has gained unanimous/widespread acceptance by the community [38,39]. This model has been empirically tested and verified by many researchers [38–41]. Figure 7.2 shows the random walk frequency offset drift of two nodes w.r.t. a reference node over time [42]. The time series for the plot in Figure 7.2 was obtained by doing real-time experiments on the GNU Radio/USRP software-defined radios.

With a model for frequency offset time evolution in hand, one could utilize a Bayesian filter (e.g., linear Kalman filter, extended Kalman filter, particle filter, etc.) to dynamically track as well as linearly predict the frequency offset over time [41,43]. The Bayesian filter would need noisy measurements of the frequency offset periodically which could be obtained blindly or using training symbols. For example, Quitin *et al.* [41] describe a blind method to estimate the frequency offset using GMSK signaling.

7.5.2   Phase synchronization

Phase synchronization is perhaps the most challenging of all. The challenge arises due to the fact that only the wrapped phase can be measured. Thus, unwrapping leads to phase ambiguities/errors [41]. The term phase offset encompasses the effects of channel phase, cumulative delay of the transmit and receive RF chains, and phase component of the oscillators involved. Figure 7.3 shows a thought experiment where multiple clocks are synchronized at time $t = 0$ w.r.t. phase, and later, they drift apart over time leading to an increase in the time deviation. Perhaps, the only viable way to ensure phase synchronization is to do an iterative

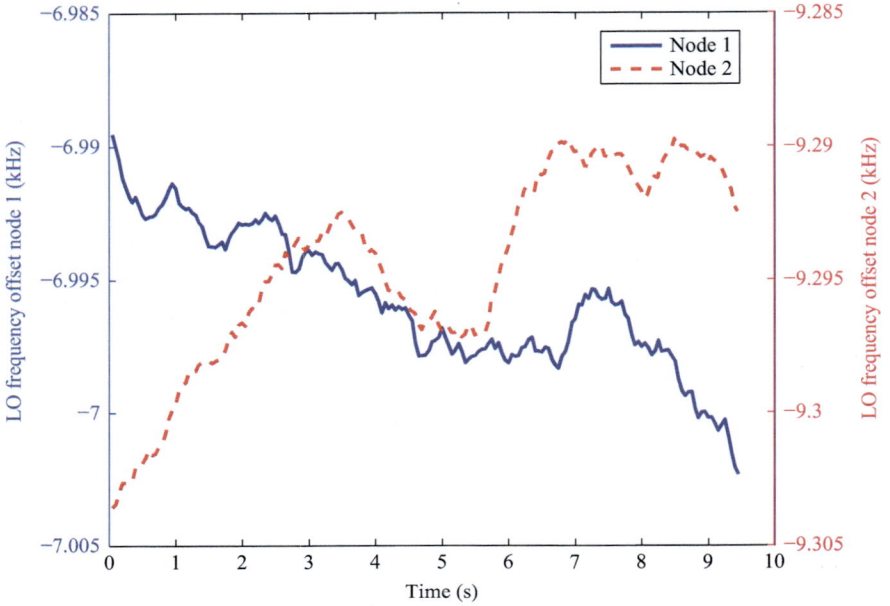

Figure 7.2   Drift of frequency offsets of two nodes w.r.t. a reference node over time [42]

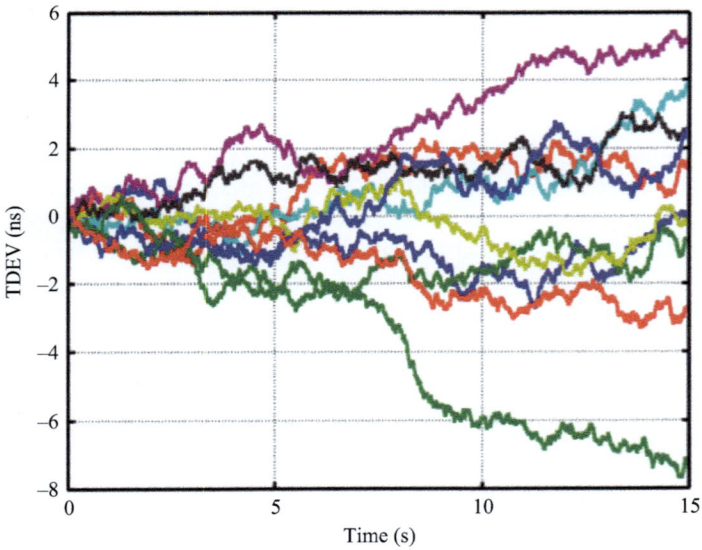

Figure 7.3   Phase drifts of the clocks over time [41]

procedure between the transmitter(s) and the receiver; the 1-bit feedback algorithm is one classical example of such an approach [41].

7.5.3   Time synchronization

Time synchronization is traditionally achieved by time of arrival (TOA) estimation, round trip time estimation, etc. TOA is in turn measured by sending a PN sequence from the transmitter while the receiver performs matched filtering and estimates the time of the peak of the output of the matched filter. Time synchronization will be challenging because the current state of the art FPGA can only resolve time intervals of duration 1 ns or higher (while slot durations for THz systems are expected to be on the order of 1 fs or so, thus, smaller by a factor of 1,000).

In short, the drift of oscillators with respect to each other calls for a systematic framework for periodic (re)-synchronization of the nodes involved in communication [41]. Figure 7.4 sketches one such framework where the nodes reserve training slots of duration T_{est} after every T_{slot} second to rectify their estimates of the frequency and phase offsets. The preamble/pilot symbols transmitted during the training interval are used by the receiver to do joint estimation of time of arrival, frequency offset, and phase offset. The noisy estimates are then passed to a Bayesian filter which generates the innovation out of the current measurement vector and updates/rectifies its predictions and estimates accordingly.

7.5.4   Network-level synchronization

The discussion so far has been limited to point-to-point synchronization. When the problem-at-hand is *network-level* synchronization, all kind of methods could be divided into two broad categories, i.e., centralized methods for network-level synchronization, distributed methods for network-level synchronization. The centralized methods come under the name master–slave architecture, while the distributed methods come under various different names, e.g., peer-to-peer (P2P) methods, consensus-based methods, etc. The pros and cons of centralized methods

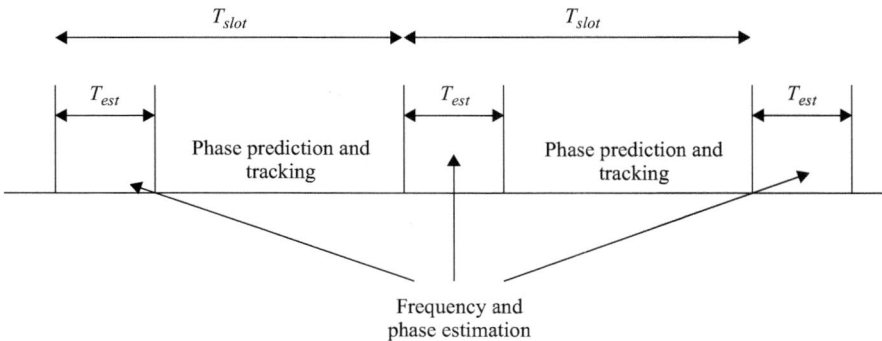

Figure 7.4   The proposed framework: drift of the clocks over time is compensated by periodic estimation of frequency and phase offsets [41]

and distributed methods are standard, i.e., centralized methods assume a central node which acts as a reference/broadcast node, single point of failure. On the other hand, in distributed methods, all the nodes participate in achieving a common goal by exchanging state info with their local neighbors.

7.5.4.1   The centralized methods: master–slave-based

The centralized methods have seen a widespread application in communication systems at microwave frequencies due to lower overhead. This is because network-level synchronization could be achieved by doing one-shot, one-way communication whereby one master node (with a very fine-grained oscillator) transmits while N nodes listen, and thereafter, synchronize themselves for the purpose of coherent transmission to some common intended receiver [42,44,45]. Figure 7.5 depicts an example scenario of a master–slave architecture.

7.5.4.2   The distributed methods: consensus-based

The distributed, peer-to-peer (P2P)-based, methods are relatively less studied in communication systems. One big class of such methods is the class of consensus-based synchronization algorithms [46,47]. Consensus problem is well-studied by the controls community in the context of multiple agents which can only exchange information with their neighbors, while they all want to achieve the same goal. Consensus-based algorithms found it hard to find their application in communication systems mainly due to the excessive overhead involved (as all the nodes exchange info with their neighbors and the algorithms are iterative in nature). Figure 7.6 sketches the block diagram/architecture of a typical consensus-based framework. Figure 7.7 plots the frequency synchronization performance of a consensus algorithm proposed by the works [46,47]. Here, N denotes the number of nodes running the consensus algorithm while α and β are the parameters of the algorithm (see [46,47], for more details).

The synchronization methods could also be classified into two different categories based upon whether the transmitter and receiver pre-share a common sequence—the training-based methods or not—the blind methods.

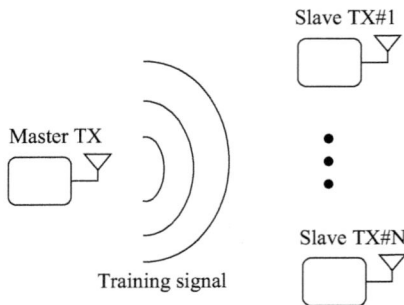

Figure 7.5   Master–slave architecture for frequency synchronization [44]

Figure 7.6   Consensus/P2P-based architecture for frequency synchronization [46,47]

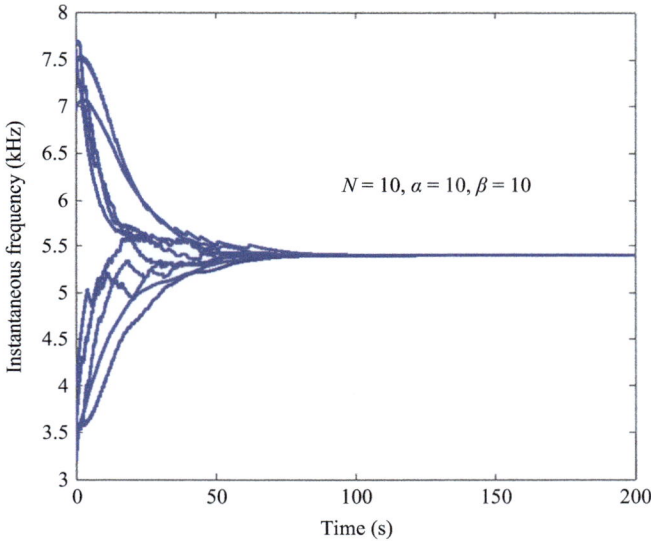

Figure 7.7   Frequency synchronization of N = 10 nodes via iterative, consensus-based algorithm [46,47]

7.5.5   Training-based methods vs. blind methods

The frequency and phase offset estimation could be blind or training data aided. The blind schemes are specific to modulation schemes being used, thus, require a priori knowledge of the modulation scheme in use. For example, Quitin *et al.* [41] utilize a blind method for frequency and phase offset estimation from the received Gaussian minimum shift keying (GMSK) signal.

7.6   Conclusion

The main takeaway messages of this chapter are as follows:

- For macro-scale communication (with a channel length of a few meters) in the THz band, CBM is to be used, while for micro-scale communication (with a channel length of a few centimeters), pulse-based modulation is used.

- The ultra-high bandwidth offered by the THz band puts a hard limit on the design of FPGAs and ADCs. Thus, there is a need to investigate mixed (analogue/digital) architectures for precoding and decoding in the THz band, in line with the literature on the millimeter (mm) wave band [4].
- The device form-factor constraints might limit the power budget at the nano-transmitter and hence the SNR at the nano-receiver. This, in turn, limits the decoding complexity at the nano-receiver as the receiver might not be able to implement, say, Viterbi algorithm with a long Trellis graph, to do MLSE for a long sequence of symbols. Thus, for M-ary modulation schemes in THz band, M might be chosen to be a small number (compared to communication at microwave frequencies where $M = 512$ is commonplace in the latest Wi-Fi and cellular standards).
- Channel coding schemes for the THz band need to have small block-length due to the limitation on computational complexity a nano-scaled device operating in the THz band could support.
- One-shot synchronization is needed to overcome frequency and phase offsets due to parts per million (ppm) specs (the manufacturing tolerances) of the oscillators. Furthermore, periodic (re-)synchronization is needed to compensate for phase and frequency drift due to aging effects and operating conditions (e.g., temperature, pressure, etc.) of the oscillators. This fact carries on to the THz band as is.
- The mathematical models that capture the oscillator dynamics at microwave frequencies [38,39] reveal that the magnitude/severity of the frequency and phase drift increases with an increase in the center frequency of operation. Therefore, with judicious extrapolation of the models in [38,39] to THz frequencies, one could expect synchronization in the THz band to be way more challenging than the synchronization at microwave frequencies (for a given performance level on a suitably chosen metric).

Additionally, a concise summary of the just-relevant points taken from the IEEE 802.15.3 standards document [1] is being provided below.

7.6.1 IEEE standard 802.15.3: executive summary

The IEEE 802.15.3 standard considers a lower THz frequency range between 252 and 325 GHz. Moreover, the IEEE standard suggests utilizing the following seven modulation schemes: BPSK, QPSK, 8-PSK, 8-APSK, 16-QAM, 64 QAM, and OOK. Additionally, the standard suggests implementing the following three-channel coding schemes: 14/15-rate LDPC (1440,1344), 11/14-rate LDPC (1440,1056), and 11/14-rate RS(240,224)-code. A lookup table that lists the achievable data rates as well as modulation and coding schemes of a THz system, given a particular MCS identifier has already been shown in Figure 7.1 for various bandwidths.

Taking the computational burden of a nano-scale transceiver as a design constraint, the standard also defines two kinds of PHY layers, THz-SC (i.e., CBM) and THz-OOK (i.e., the pulse-based modulation). The two PHY layers (THz-SC and

THz-OOK) enable data rates up to 100 Gb/s using eight different bandwidths between 2 and 70 GHz. The THz-OOK PHY is meant for the design of the cost-effective nano-scale transceivers that require low power, low complexity, and simple design. For applications using this PHY, transmission ranges of a few tens of centimeters are targeted. The THz-OOK PHY supports a single modulation scheme, OOK, and three FEC schemes. The RS code is mandatory and allows simple decoding without soft-decision information. The LDPC codes with rates of 14/15 and 11/15 are optional and allow the use of soft-decision information. On the other hand, The THz-SC PHY is designed for extremely high bit rates of about 100 Gb/s while selecting from a range of bandwidths. Thus, with THz-SC, quite high data rates are achievable, depending on the MCS chosen and the bandwidth. The THz-SC PHY supports a wide range of modulations: $\pi/2$ BPSK, $\pi/2$ QPSK, $\pi/2$ 8-PSK, $\pi/2$ 8-APSK, 16-QAM, and 64-QAM. The FEC consists of two LDPC codes with rates of 14/15 and 11/15.

References

[1]   IEEE standard for high data rate wireless multi-media networks—amendment 2: 100 Gb/s wireless switched point-to-point physical layer. *IEEE Std 802.15.3d-2017 (Amendment to IEEE Std 802.15.3-2016 as amended by IEEE Std 802.15.3e-2017)*, pp. 1–55, Oct 2017.

[2]   J. M. Jornet and I. F. Akyildiz. Femtosecond-long pulse-based modulation for terahertz band communication in nanonetworks. *IEEE Transactions on Communications*, 62(5):1742–1754, 2014.

[3]   R. G. Cid-Fuentes, J. M. Jornet, I. F. Akyildiz, and E. Alarcon. A receiver architecture for pulse-based electromagnetic nanonetworks in the terahertz band. In *2012 IEEE International Conference on Communications (ICC)*, pp. 4937–4942, June 2012.

[4]   A. Alkhateeb, O. El Ayach, G. Leus, and R. W. Heath. Channel estimation and hybrid precoding for millimeter wave cellular systems. *IEEE Journal of Selected Topics in Signal Processing*, 8(5):831–846, 2014.

[5]   A. K. Vavouris, F. D. Dervisi, V. K. Papanikolaou, and G. K. Karagiannidis. An energy efficient modulation scheme for body-centric nano-communications in the THz band. In *2018 7th International Conference on Modern Circuits and Systems Technologies (MOCAST)*, pp. 1–4, May 2018.

[6]   R. Zhang, K. Yang, Q. H. Abbasi, K. A. Qaraqe, and A. Alomainy. Analytical characterisation of the terahertz in-vivo nano-network in the presence of interference based on THz-OOK communication scheme. *IEEE Access*, 5:10172–10181, 2017.

[7]   F. Moshir and S. Singh. Modulation and rate adaptation algorithms for terahertz channels. *Nano Communication Networks*, 10:38–50, 2016 (Terahertz Communications).

[8]   M. O. Iqbal, M. M. Ur Rahman, M. A. Imran, A. Alomainy, and Q. H. Abbasi. Modulation mode detection and classification for in vivo nano-scale

communication systems operating in terahertz band. *IEEE Transactions on NanoBioscience*, 18(1):10–17, 2019.

[9] A. Hirata, T. Kosugi, H. Takahashi, *et al*. 120-GHz-band millimeter-wave photonic wireless link for 10-Gb/s data transmission. *IEEE Transactions on Microwave Theory and Techniques*, 54(5):1937–1944, 2006.

[10] H.-J. Song, K. Ajito, A. Hirata, *et al*. Multi-gigabit wireless data transmission at over 200-GHz. In *2009 34th International Conference on Infrared, Millimeter, and Terahertz Waves*, pp. 1–2. IEEE, 2009.

[11] G. Ducournau, P. Szriftgiser, D. Bacquet, *et al*. Optically power supplied Gbit/s wireless hotspot using 1.55 μm THz photomixer and heterodyne detection at 200 GHz. *Electronics Letters*, 46(19):1349–1351, 2010.

[12] A. Hirata, T. Kosugi, H. Takahashi, *et al*. 5.8-km 10-Gbps data transmission over a 120-GHz-band wireless link. In *2010 IEEE International Conference on Wireless Information Technology and Systems*, pp. 1–4. IEEE, 2010.

[13] C. Jastrow, S. Priebe, B. Spitschan, *et al*. Wireless digital data transmission at 300 GHz. *Electronics Letters*, 46(9):661–663, 2010.

[14] L. Moeller, J. Federici, and K. Su. THz wireless communications: 2.5 Gb/s error-free transmission at 625 GHz using a narrow-bandwidth 1 mW THz source. In *2011 XXXth URSI General Assembly and Scientific Symposium*, pp. 1–4. IEEE, 2011.

[15] I. Kallfass, J. Antes, T. Schneider, *et al*. All active MMIC-based wireless communication at 220 GHz. *IEEE Transactions on Terahertz Science and Technology*, 1(2):477–487, 2011.

[16] J. Antes, J. Reichart, D. Lopez-Diaz, *et al*. System concept and implementation of a MMW wireless link providing data rates up to 25 Gbit/s. In *2011 IEEE International Conference on Microwaves, Communications, Antennas and Electronic Systems (COMCAS 2011)*, pp. 1–4. IEEE, 2011.

[17] X. Pang, A. Caballero, A. Dogadaev, *et al*. 100 Gbit/s hybrid optical fiber-wireless link in the W-band (75–110 GHz). *Optics Express*, 19(25):24944–24949, 2011.

[18] B. Zhang, Y.-Z. Xiong, L. Wang, and S. Hu. A switch-based ASK modulator for 10 Gbps 135 GHz communication by 0.13 micro-meter MOSFET. *IEEE Microwave and Wireless Components Letters*, 22(8):415–417, 2012.

[19] H.-J. Song, K. Ajito, Y. Muramoto, A. Wakatsuki, T. Nagatsuma, and N. Kukutsu. 24 Gbit/s data transmission in 300 GHz band for future terahertz communications. *Electronics Letters*, 48(15):953–954, 2012.

[20] M. J. Fice, E. Rouvalis, F. Van Dijk, *et al*. 146-GHz millimeter-wave radio-over-fiber photonic wireless transmission system. *Optics Express*, 20(2):1769–1774, 2012.

[21] J. Antes, S. Koenig, A. Leuther, *et al*. 220 GHz wireless data transmission experiments up to 30 gbit/s. In *2012 IEEE/MTT-S International Microwave Symposium Digest*, pp. 1–3. IEEE, 2012.

[22] K. Ishigaki, M. Shiraishi, S. Suzuki, M. Asada, N. Nishiyama, and S. Arai. Direct intensity modulation and wireless data transmission characteristics

of terahertz-oscillating resonant tunnelling diodes. *Electronics Letters*, 48(10):582–583, 2012.

[23] C. Wang, C. Lin, Q. Chen, B. Lu, X. Deng, and J. Zhang. A 10-Gbit/s wireless communication link using 16-QAM modulation in 140-GHz band. *IEEE Transactions on Microwave Theory and Techniques*, 61(7):2737–2746, 2013.

[24] J. Antes, S. Koenig, D. Lopez-Diaz, *et al.* Transmission of an 8-PSK modulated 30 Gbit/s signal using an MMIC-based 240 GHz wireless link. In *2013 IEEE MTT-S International Microwave Symposium Digest (MTT)*, pp. 1–3. IEEE, 2013.

[25] S. Sarkozy, M. Vukovic, J. G. Padilla, *et al.* Demonstration of a g-band transceiver for future space crosslinks. *IEEE Transactions on Terahertz Science and Technology*, 3(5):675–681, 2013.

[26] B. Lu, W. Huang, C. Lin, and C. Wang. A 16QAM modulation based 3 Gbps wireless communication demonstration system at 0.34 THz band. In *2013 38th International Conference on Infrared, Millimeter, and Terahertz Waves (IRMMW-THz)*, pp. 1–2. IEEE, 2013.

[27] H.-J. Song, J.-Y. Kim, K. Ajito, M. Yaita, and N. Kukutsu. Fully integrated ASK receiver MMIC for terahertz communications at 300 GHz. *IEEE Transactions on Terahertz Science and Technology*, 3(4):445–452, 2013.

[28] T. Nagatsuma, S. Horiguchi, Y. Minamikata, *et al.* Terahertz wireless communications based on photonics technologies. *Optics Express*, 21(20): 23736–23747, 2013.

[29] S. Koenig, D. Lopez-Diaz, J. Antes, *et al.* Wireless sub-THz communication system with high data rate. *Nature Photonics*, 7(12):977, 2013.

[30] G. Ducournau, P. Szriftgiser, A. Beck, *et al.* Ultrawide-bandwidth single-channel 0.4-THz wireless link combining broadband quasi-optic photomixer and coherent detection. *IEEE Transactions on Terahertz Science and Technology*, 4(3):328–337, 2014.

[31] J. M. Jornet and I. F. Akyildiz. Low-weight channel coding for interference mitigation in electromagnetic nanonetworks in the terahertz band. In *2011 IEEE International Conference on Communications (ICC)*, pp. 1–6, June 2011.

[32] S. T. Chung and A. J. Goldsmith. Degrees of freedom in adaptive modulation: A unified view. *IEEE Transactions on Communications*, 49(9):1561–1571, 2001.

[33] A. Moldovan, S. Kisseleff, I. F. Akyildiz, and W. H. Gerstacker. Data rate maximization for terahertz communication systems using finite alphabets. In *2016 IEEE International Conference on Communications (ICC)*, pp. 1–7. IEEE, 2016.

[34] H. Sarieddeen, M.-S. Alouini, and T. Y. Al-Naffouri. Terahertz-band ultra-massive spatial modulation MIMO. Arxiv preprint, 2019.

[35] A. Gupta, M. Medley, and J. M. Jornet. Joint synchronization and symbol detection design for pulse-based communications in the THz band. In *Global Communications Conference (GLOBECOM)*, pp. 1–7. IEEE, 2015.

[36]   Q. Xia, Z. Hossain, M. Medley, and J. M. Jornet. A link-layer synchronization and medium access control protocol for terahertz-band communication networks. In *2015 IEEE Global Communications Conference (GLOBECOM)*, pp. 1–7, Dec 2015.

[37]   B. T. Bulcha, J. L Hesler, V. Drakinskiy, *et al.* Design and characterization of 1.8–3.2 THz Schottky-based harmonic mixers. *IEEE Transactions on Terahertz Science and Technology*, 6(5):737–746, 2016.

[38]   C. Zucca and P. Tavella. The clock model and its relationship with the Allan and related variances. *IEEE Transactions on Ultrasonics, Ferroelectrics and Frequency Control*, 52(2):289–296, 2005.

[39]   L. Galleani. A tutorial on the two-state model of the atomic clock noise. *Metrologia*, 45(6):S175, 2008.

[40]   M. M. Ur Rahman. Distributed beamforming and nullforming: Frequency synchronization techniques, phase control algorithms, and proof-of-concept. University of Iowa, 2013.

[41]   F. Quitin, M. M. Ur Rahman, R. Mudumbai, and U. Madhow. A scalable architecture for distributed transmit beamforming with commodity radios: Design and proof of concept. *IEEE Transactions on Wireless Communications*, 12(3):1418–1428, 2013.

[42]   F. Quitin, M. M. Ur Rahman, R. Mudumbai, and U. Madhow. Distributed beamforming with software-defined radios: Frequency synchronization and digital feedback. In *2012 IEEE Global Communications Conference (GLOBECOM)*, pp. 4787–4792. IEEE, 2012.

[43]   M. M. U. Rahman, A. Yasmeen, and J. Gross. PHY layer authentication via drifting oscillators. In *2014 IEEE Global Communications Conference*, pp. 716–721, Dec 2014.

[44]   M. M. Rahman, H. E. Baidoo-Williams, R. Mudumbai, and S. Dasgupta. Fully wireless implementation of distributed beamforming on a software-defined radio platform. In *2012 ACM/IEEE 11th International Conference on Information Processing in Sensor Networks (IPSN)*, pp. 305–315. IEEE, 2012.

[45]   F. Quitin, U. Madhow, M. M. Ur Rahman, and R. Mudumbai. Demonstrating distributed transmit beamforming with software-defined radios. In *2012 IEEE International Symposium on a World of Wireless, Mobile and Multimedia Networks (WoWMoM)*, pp. 1–3. IEEE, 2012.

[46]   M. M. Ur Rahman, R. Mudumbai, and S. Dasgupta. Consensus based carrier synchronization in a two node network. *IFAC Proceedings Volumes*, 44(1): 10038–10043, 2011.

[47]   M. M. Rahman, S. Dasgupta, and R. Mudumbai. A distributed consensus approach to synchronization of RF signals. In *2012 IEEE Statistical Signal Processing Workshop (SSP)*, pp. 281–284. IEEE, 2012.

Chapter 8

Routing protocols for nano-electromagnetic communication networks

Xin-Wei Yao[1], Ye-Chen-Ge Wu[1], Chao-Chao Wang[1] and Wei Huang[1]

Nano-electromagnetic communication networks, namely, wireless nano networks (WNNs), are wireless communication networks composed of interacting nanonodes (sizes ranging from a few hundred cubic nanometers to micrometers). The extremely limited capabilities and resources of nanonodes, as well as the severe path loss of terahertz band communication in WNNs, represent a challenge to the communication distance among nanonodes and the overall network performance. Therefore, appropriate routing protocols are necessary for guaranteeing multihop communication in WNNs. In this chapter, the existing routing protocols for WNNs are comprehensively analyzed and classified into three categories: limit-flood-area-based routing protocols, dynamic-infrastructure-based (DIF-based) routing protocols, and single-path-based routing protocols. Based on the peculiarities of WNNs, especially the constrained resources and limited energy supply, the features of each protocol are presented through a detailed comparison. Finally, by integrating the features of WNNs and the problems of existing routing technologies, we present our views on the future research directions of routing techniques in WNNs.

8.1 Introduction

Nanotechnology is the technology that enables the creation and production of microdevices, such as nanodevices. Nanosensors, one of the earliest nanodevices, have been widely researched and developed [1–4]. In general, devices in the scale ranging from a few hundred nanometers to micrometers have unique characteristics, including new functionalities and limitations [5]. Although nanonodes have limited abilities to sense, compute, memorize, and manage energy, through interconnecting with each other, a large number of nanonodes can overcome the individual limitations and can benefit from collaborative efforts in a nanonetwork.

---

[1]College of Computer Science and Technology, Zhejiang University of Technology, Hangzhou, China

Since nanosensors are able to sense new events at the nanoscale, nanonetworks can potentially be widely utilized in many unprecedented areas, such as the biomedical field (e.g., intrabody health monitoring and drug delivery systems), industrial field, environmental field (e.g., air pollution control), and military field (e.g., surveillance against chemical attacks at the nanoscale) [5,6]. Due to the potential applications of nanonetworks, and with the development of nanotechnology and materials, research on nanonetworks has been abundant in recent years.

Wireless sensor networks (WSNs) consist of small autonomous devices called nodes or motes that harvest information such as temperature, pressure, or vibration from their physical environment [7]. WNN is a special type of WSN whose nodes are nanoscale and have more restrictions than the nodes in WSNs. There are four nanoscale communication technologies for nanonetworks: electromagnetic, acoustic, nanomechanical, and molecular. Since acoustic waves are strongly absorbed by the human body, acoustic communication is not suitable for WNNs [8]. Nanomechanical communication requires nanonodes to be physically connected to each other. Therefore, it cannot be used directly in nanonetworks. Electromagnetic communication is defined as the transmission and reception of electromagnetic radiation from components based on novel nanomaterials [9]. However, electromagnetic communication nanonetworks did not receive substantial attention from researchers in the early years because electronic nanocomponents of nanonodes had not been manufactured. Therefore, many researchers mainly focused on molecular communication in the past few decades [10].

Fortunately, recent advancements in molecular and carbon electronics have opened the door to a new generation of electronic nanocomponents, such as nanobatteries, nanomemories, nanoscale logical circuitry, and even nanoantennas [11]. Electromagnetic communication was eventually developed with the advancements in graphene-based electronics [12]. Since one-micrometer-long graphene-based antennas were demonstrated to effectively radiate electromagnetic waves in the terahertz (THz) band (0.1–10.0 THz) in 2010 [12], nanodevices have achieved high transmission rates over short distances while communicating [13,14]. THz band communication has been envisioned as a key wireless technology for providing unprecedented high data rates [15]. With the development of nanocomponents, electromagnetic communication and molecular communication are recommended as the most promising communication methods for nanonetworks [16]. Although nanonodes are capable of electromagnetic communication, the resulting nanonetworks still have many limitations:

- Nanonodes have extremely weak processing capability and small memory units due to their nanoscale size.
- A nanonode energy-harvesting process is required to overcome the limited energy supply. Some traditional studies on the energy-harvesting process in WSNs have been presented, but the particular characteristics of the nanoscale energy-harvesting process were not considered [17]. In this regard, energy-harvesting technologies at the nanoscale, which convert different forms of

energy, e.g., vibrational, fluidic, electromagnetic, or acoustic, into electrical energy, have recently been proposed [18–20]. Moreover, in order to achieve a perpetual nanonetwork, the tradeoff between energy harvesting and consumption needs to be considered [21].

- A nanonode in WNNs has very small nanobatteries as a result of nanoscale components [22]. The limited energy storage of nanonodes and the peculiarities of the energy-harvesting process, which result in the fluctuation of the remaining energy of the nanonodes over time.
- Due to the high density of nanonodes in WNNs, uncontrolled broadcast scenarios will lead to severe redundancy and collisions [23].
- Due to the severe molecular absorption and limited transmission power, THz band communications in WNNs will suffer significant path loss and affect the overall network performance [13]. In a study of the achievable throughput of terahertz, energy-harvesting nanonetworks found that the throughput decreases with the molecular absorption loss coefficient, which indicates some frequency sub-bands with high absorption loss should be avoided [24].

In light of the above-listed limitations, appropriate protocols for each layer of the nanonetwork must be designed. The routing protocol is the process of selecting communication paths in a network and is a crucial building block in a network protocol stack. As in WSNs, the routing protocol also plays a vital role in the efficient delivery of packets to their respective destinations under given energy and complexity constraints. In WSNs, routing protocols usually aim to minimize energy consumption and maximize the lifetime of the network [25], but routing protocols with good performance generally have high computational complexity. Therefore, traditional routing protocols developed for WSNs are not directly suitable for WNNs, mainly due to the limited processing capability and energy storage of nanonodes. Early transmission approaches in nanonetworks were based on flooding, in which each node blindly rebroadcasts the packet the node receives for the first time [5]. Although flood-based routing protocols can achieve high robustness, the high density of nanonodes leads to serious redundancy and conflicts [23]. In this light, better routing protocols are required for more efficient transmission in nanonetworks. Therefore, we conduct this chapter of the routing protocols of nanonetworks.

This chapter aims to perform a comprehensive and detailed study of routing protocols in WNNs. Since a routing protocol may have specific characteristics that depend on the application and network architecture [26], the corresponding classification of nanonetworks is presented first. Then, the existing routing protocols are compared from different perspectives, with an analysis of the advantages and disadvantages of each routing protocol. With the aforementioned comprehensive analysis, some recommendations are made for the development of WNNs routing protocols.

The rest of this chapter is organized as follows. In Section 8.2, the classifications of WSNs and WNNs are briefly presented and compared. In Section 8.3, different WNN routing protocols are reviewed in detail. Finally, conclusions are drawn in Section 8.4, with a brief discussion of future work.

8.2   Classification of routing protocols in WNNs

To better describe the routing protocols in WNNs, WNNs are classified according to the appropriate principles. Since WNNs and WSNs have some similar characteristics, many rules and experiences with classification in WSNs can be adopted for WNNs.

Some studies on routing protocols for WNNs have been reported. However, the traditional classification principles cannot be used directly for the routing protocols in WNNs for the following reasons:

- The communication model is the operation of the routing protocol to route data. Several communication models exist for WSNs, such as query, negotiation, coherent, and non-coherent. However, the existing routing protocols in WNNs do not consider these communication models. Therefore, the communication models are not suitable for routing protocol classification in WNNs.
- The topology maintenance strategy is based on the way in which the nodes maintain topology information. However, in WNNs, the nanonodes do not have sufficient memory to store topology information. Therefore, the topology maintenance strategy cannot be used in the classification of WNNs.
- Reliable routing is where the routing protocols aim to satisfy QoS metrics. However, most of the existing routing protocols do not consider QoS. Therefore, reliable routing is not recommended in the classification of WNNs.
- In WNNs, the nanonodes in most of the routing protocols are deployed randomly, and the routing protocols are not designed for the specific deployment strategies of WNNs; Thus, the deployment strategy cannot be used in the classification of WNNs.
- In WNNs, the nanonodes in all the routing protocols are assumed to be static; thus, the mobility of nanonodes cannot be used in the classification of WNNs.

Since the research of WNN routing protocols is still in infancy, most of the published routing protocols are based on flooding or single path, which reduces the energy consumption of peer-to-peer communication by reducing the number of nanonodes participating in forwarding. Therefore, in this chapter, the existing routing protocols are classified into limited-flood-area-based routing protocols, DIF-based routing protocols and single-path-based routing protocols according to the routing strategy, as shown in Figure 8.1. It will inevitably conduct in-depth research on topology maintenance and network reliability when the research on the reduction of energy consumption in WNN routing protocols is mature.

8.3   Routing protocols in WNNs

Due to the limitations of nanonodes, that is, low processing capability, small memory units, and high density, routing in WNNs is more challenging than that in WSNs. Therefore, the existing studies on routing protocols in WNNs are primarily concerned with reducing complexity and energy consumption. Few routing

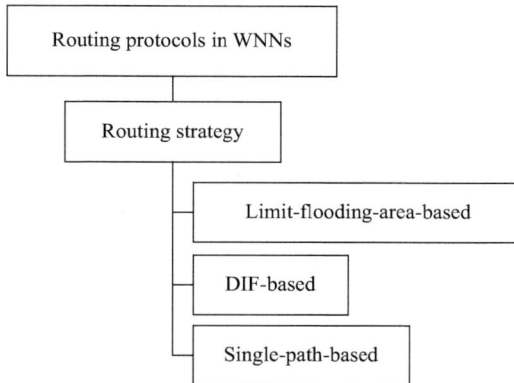

Figure 8.1   The classification of WNN routing protocols

protocols that consider the inherent features of WNNs along with the architecture requirements have been proposed, for example, the limit-flood-area-based routing protocols RADAR, CORONA, SLR, etc., the DIF-based routing protocols LSDD, DEROUS, etc., and the single-path-based routing protocols MHTD, ECR, EEMR, TEFoward, etc. The rest of this section will introduce the limit-flood-area-based routing protocols, DIF-based routing protocols, and single-path-based routing protocols.

8.3.1   Limit-flood-area-based routing protocols

Flooding routes allow all reachable nanonodes to forward data, in which each transmission will cause many redundant nanonodes to consume energy for forwarding. Limiting flooding area can greatly reduce the number of nanonodes involved in forwarding, thereby reducing energy consumption.

8.3.1.1   RADAR routing

In RADAR routing [27], nanonodes are evenly distributed in a given circular area. An entity placed in the center of the circular area rotates regularly and emits radiation at a certain angle, as shown in Figure 8.2. The nanonodes in the region of radiation are in the on state, while the other nodes are in the off state. The nanonodes in the off state do not consume energy. Packets are transmitted by flooding the region with radiation; hence, the number of packets transmitted in the nano-network is reduced. However, RADAR routing may suffer from packet loss since the receiver may be in the off state. Therefore, the author defined the parameter k as the probability of a packet reaching its destination in this routing protocol. The number of received packets m and the number of transmitted packets n are recorded to calculate the value of this parameter. The ratio between m and n is the value of k.

In addition to possible packet loss, RADAR routing has another obvious drawback: the farther from the central entity, the more nodes in the one state. As a result, collisions caused by a large number of packets may occur. This shortcoming

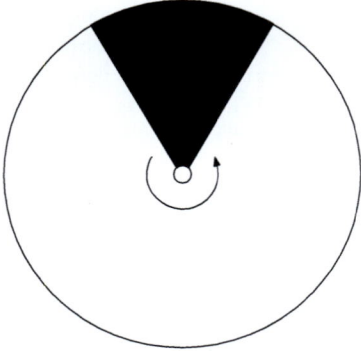

Figure 8.2   An example of RADAR routing

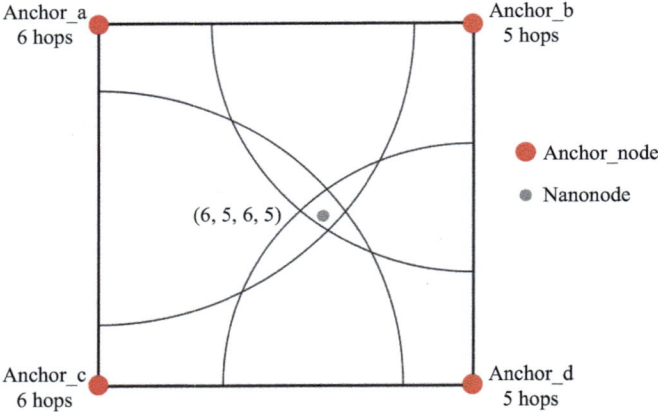

Figure 8.3   Coordination system setup [23]

becomes more serious in large-scale nanonetworks. Therefore, RADAR routing may not be suitable in large-scale nanonetworks.

8.3.1.2   CORONA

The routing protocol CORONA assigns addresses in nanonetworks in the form of a coordinate system [23]. In this protocol, each nanonode derives its own coordinates dynamically in a setup process.

In the CORONA protocol, a large set of nanonodes are assumed to be uniformly placed in a rectangular area. Four anchor nodes are placed in the four corners of the rectangular area. In the setup phase, each anchor node sends a packet in the sequence. All nanonodes set the hop counts from four anchor nodes as their respective coordinates, as shown in Figure 8.3.

After the coordinate setting, when nanonode A wants to send a packet to another nanonode B, other nanonodes whose coordinates are between the coordinates of nanonodes A and B retransmit the packet. As shown in Figure 8.4, the packet is retransmitted by flooding the red area. However, the two facing anchor nodes cannot be selected for routing; otherwise, the transmission may not be completed, as shown in Figure 8.5.

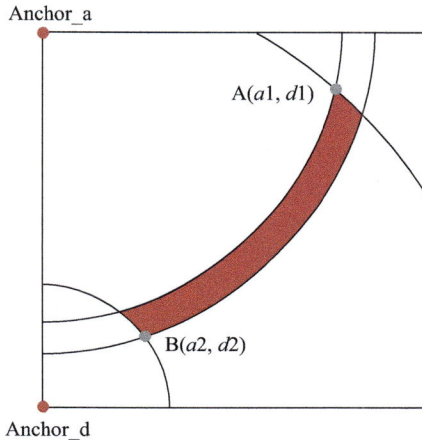

Figure 8.4   *Using anchor_a and anchor_d for routing; the red area is the area for which the coordinates are between those of node A and node B*

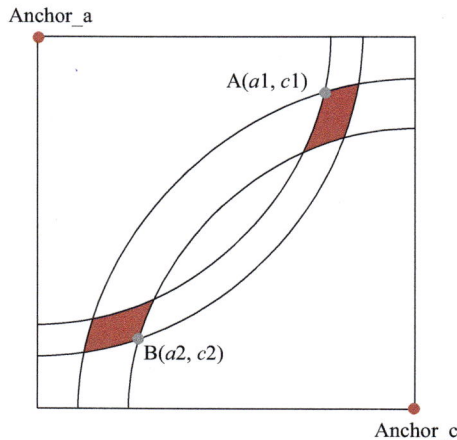

Figure 8.5   *When using facing anchor points for routing, the transmission may not be completed*

The hop counts from a nanonode to the four anchor nodes are defined as the coordinates in CORONA, and the flooding is limited to an arc-shaped path between any two nanonodes. These paths have been proven to be effective or point-to-point communication, improving the energy efficiency in the highly restricted nano-environment [23].

8.3.1.3   Stateless linear routing

Stateless linear routing (SLR) is a coordinate-based routing protocol designed for 3D nanonetworks [28,29] in which the data are routed in a straight line. The coordinate system of SLR is extended from CORONA and assigns 3D coordinates to nanonodes. All nanonodes are assumed to be placed in a cubic space, where eight anchor nodes are placed at the vertexes, as shown in Figure 8.6.

In the setup phase, the anchor nodes send setup packets in sequence. The hop counts from the nanonodes to the anchor nodes are called the distances. Three anchor nodes in the same face can uniquely identify a zone [28]; thus, all nanonodes in the space set three anchor distances as their coordinates.

In the calculation phase, the SLR protocol deduces whether a nanonode is located on the straight line from the sender to the receiver. If so, the node will retransmit data. The algorithm for determining whether a nanonode is on the straight line between the sender and the receiver is as follows.

The coordinates of the sender and receiver are assumed to be $(a1, b1, c1)$ and $(a2, b2, c2)$, respectively. If a nanonode P, whose coordinates are (a, b, c), is on the straight line between the sender and the receiver, (8.1) must be satisfied:

$$\frac{a - a1}{a2 - a1} = \frac{b - b1}{b2 - b1} = \frac{c - c1}{c2 - c1} \tag{8.1}$$

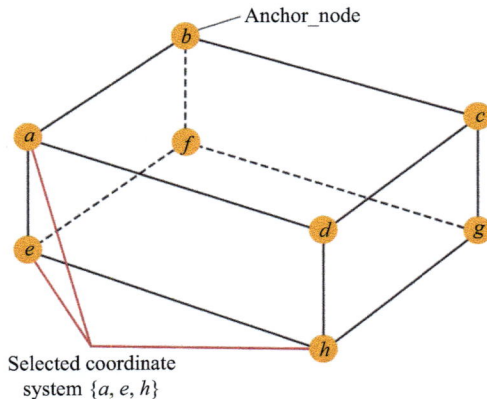

Figure 8.6   Overview of the 3D nanonetwork space [29]

Since the coordinates are composed of hop counts, which are integers, (8.1) can be converted to (8.2) as follows:

$$\begin{cases} (a - a1)(b2 - b1) - (b - b1)(a2 - a1) = 0 \\ (a - a1)(c2 - c1) - (c - c1)(a2 - a1) = 0 \end{cases} \tag{8.2}$$

Then, $\Delta^x(a, b)$ and $\Delta^y(a, c)$ are defined as

$$\begin{aligned} \Delta^x(a, b) &= (a - a1)(b2 - b1) - (b - b1)(a2 - a1) \\ \Delta^y(a, c) &= (a - a1)(c2 - c1) - (c - c1)(a2 - a1) \end{aligned} \tag{8.3}$$

According to (8.2), if nanonode P is on the straight line between the sender and the receiver, the values of $\Delta^x(a, b)$ and $\Delta^y(a, c)$ are 0. To easily calculate the value and to ensure multipath and robustness, the deviation value m is defined to allow the nanonode P to deviate slightly from the straight line. When altering a, b, and c, the quantities are defined as follows:

$$\begin{aligned} \Delta^x_a(a, b) &= \Delta^x(a - m, b) \\ \Delta^x_b(a, b) &= \Delta^x(a, b - m) \\ \Delta^x_{ab}(a, b) &= \Delta^x(a - m, b - m) \end{aligned} \tag{8.4}$$

$$\begin{aligned} \Delta^y_a(a, c) &= \Delta^y(a - m, c) \\ \Delta^y_b(a, c) &= \Delta^y(a, c - m) \\ \Delta^y_{ac}(a, c) &= \Delta^y(a - m, c - m) \end{aligned} \tag{8.5}$$

Now, we need to check only whether $\Delta^x(a, b)$ and $\Delta^y(a, c)$ undergo a sign change according to (8.6). If so, nanonode P is within the allowable deviation and will retransmit the data. Then, all the retransmitters form a wedge shape, where the deviation value m regulates the "width" of the transmission path:

$$\begin{cases} \Delta^x \cdot \Delta^x_a \leq 0 \text{ or } \Delta^x \cdot \Delta^x_b \leq 0 \text{ or } \Delta^x \cdot \Delta^x_{ab} \leq 0 \\ \Delta^y \cdot \Delta^y_a \leq 0 \text{ or } \Delta^y \cdot \Delta^y_c \leq 0 \text{ or } \Delta^y \cdot \Delta^y_{ac} \leq 0 \end{cases} \tag{8.6}$$

However, the anchor-node distances used in the coordinates must be analyzed in SLR. A set of three anchor-node distances that can identify a zone are referred to as a viewport. Moreover, a viewport selection optimization process is considered in [29] to enable the sender to select the best viewports.

The goal of SLR is the same as that of CORONA, i.e., to limit the flooding area.

8.3.2   *DIF-based routing protocols*

The dynamic infrastructure (DIF) scheme can effectively reduce energy consumption based on flood routing and even control the direction of data transmission. The specific method is introduced in the following two routing protocols LSDD and DEROUS.

8.3.2.1   LSDD

In this routing protocol, a flood-based communication paradigm is adopted, in which highly scalable node communication is offered [30]. This flood-based routing protocol leads to simple node architecture and low manufacturing cost for the nanonodes.

In LSDD, one nanonode placed in the center senses data and seeks to transmit the data to an external entity. While data packets are transmitted in the nanonetwork, all nanonodes operate their own packet-receive statistics and are self-classified as retransmitters or passive auditors [31]. Three types of packet-receive statistics exist:

> "PARITY_ERROR:" The packet has been received but failed the integrity checks.
> "DUPLICATION_ERROR:" The packet has been successfully received and has passed the integrity checks, but the packet was received repeatedly.
> "RECEPTION_SUCCESS:" The packet has been successfully received and has passed the integrity checks. In addition, the packet was received for the first time.

A sequence formed by these statistics is processed by a simplification of the Misra–Gries algorithm [32]. A nanonode's ability to serve as a retransmitter is assessed by finding the most frequent item in the sequence. The retransmitter then blindly forwards all incoming packets, whereas the auditors do not participate in the retransmission process. This self-classification process has been shown to exhibit several good properties [33]. An example is given in Figure 8.7, where the central nanonode senses data and seeks to transmit the data to an external entity. Packets are forced to follow straight paths to reduce the delivery time, and because of the straight-path transmission, the location of a sensing event can be pinpointed

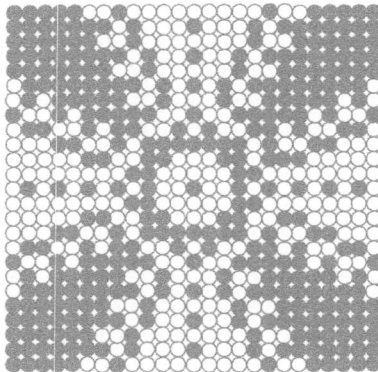

Figure 8.7   An example of nanonode classification. Nanonodes in white serve as retransmitters, whereas gray nanonodes are passive auditors [31]

with the ray-like formation in random topologies by performing power measurements around the nanonetwork area [33].

Although this routing protocol achieves high robustness and energy efficiency with a simple network architecture by considering the deployment cost and network scalability, the protocol can reduce the packet transmission rate.

8.3.2.2 DEROUS

As in RADAR routing, nanonodes are evenly distributed in a circular area in the DEROUS (the proposed deployable routing system) routing protocol [34]. A nanonode in the center of the circular area is set as the beacon node, which assigns addresses to the other nanonodes in the nanonetwork. The beacon node periodically transmits setup packets, and all other nanonodes set the hop counts as the radius, as depicted in Figure 8.8.

During the setup phase, as in LSDD, each nanonode records its packet reception status in the form of success or failure and saves the hop counts from the beacon node. Based on these simple integer records, each nanonode can determine whether its reception quality is acceptable and therefore become "infrastructure" or "user." At the same time, the hop counts from the beacon node can be considered to be the radius to the center of the circle. This process is lightweight and virtually instantaneous given that it can be completed within three beacon packet emissions [31].

The simulation results of the process show that the pattern of the infrastructure nodes is predictable and depends on the transmission radius of the nanonode [33]. A larger transmission radius results in the retransmitters forming a circular pattern centered on the beacon node, while a smaller transmission radius results in the

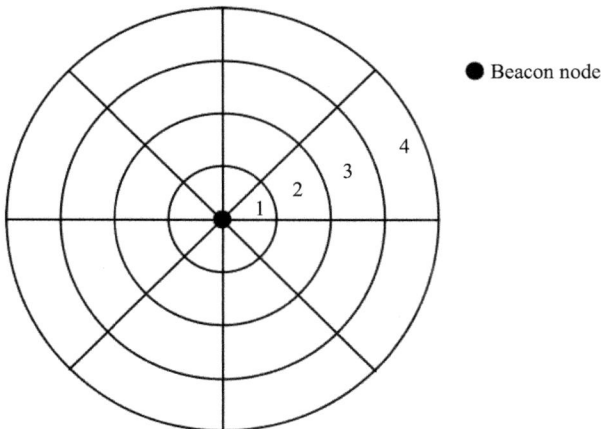

Figure 8.8   The basic principles of the DEROUS system. DEROUS deploys rings and sectors based on the classification results generated by different radii. Packets can then be routed over the combined system, driven by their distance in hops (N_HOPS) from the beacon O [35]

retransmitters forming a radial line. Thus, the direction of diffusion is divided into angular diffusion and radial diffusion, depending on the transmission radius of the nanonode. The transmission radius can be adjusted by changing the transmission power.

When a nanonode sends a packet, the nanonode whose radius is between the sender's and the receiver's will retransmit the packet in the radial direction with low power. Another situation occurs when a node's radius is the same as that of the sender's or receiver's: the nanonode will retransmit the packet in the angular direction with normal power.

DEROUS dynamically forms circular and radial packet routing paths that effectively serve peer-to-peer communication requirements while considerably limiting redundant transmissions and ensuring a good degree of path multiplicity. DEROUS limits data transmission to a circular area and is more energy efficient than LSDD.

8.3.3   Single-path-based routing protocols

The energy consumption of single-path-based routing is always lower than that of flood-based routing, but it may lead to high latency and high packet loss rate. Therefore, in single-path-based routing, an optimized transmission path needs to be found to reduce power consumption, delay, and packet loss rate. In this subsection, we introduce four single-path-based routing protocols for WNNs, namely, MHTD, EEMR, ECR, and TEFoward.

8.3.3.1   Multihop transmission decision

Pierobon *et al.* proposed a routing framework for nanonetworks in [6]. This routing framework aims to ensure that the network lifetime reaches infinity while ensuring sufficient network throughput. In the routing framework, a multihop transmission decision (MHTD) routing protocol is used in the hierarchical architecture nanonetwork. A nanocontroller with more computing resources than the nanosensors acts as the cluster head in a cluster. The MHTD protocol calculates the probability that the average energy consumption is lower when the multihop transmission is adopted instead of single-hop transmission. If multihop transmission is adopted, then the transmission power of each nanonode is tuned to optimize the one-hop distance and the throughput of the nanonetwork.

Communication in the routing framework follows the dynamic time division multiple access (TDMA) scheduling. Considering the process of data transmission, the structure of the time frame comprises four subframes: downlink (DL), uplink (UL), multihop (MH), and random access (RA). Based on the TDMA scheduling, the MHTD protocol is used when nanosensors want to transmit data. The communication process includes the following steps, as shown in Figure 8.9:

1. A nanosensor n directly transmits a request to the nanocontroller in the RA subframe.
2. When the nanocontroller receives the request, the nanocontroller calculates the probability of saving energy through multihop transmission PES(n).

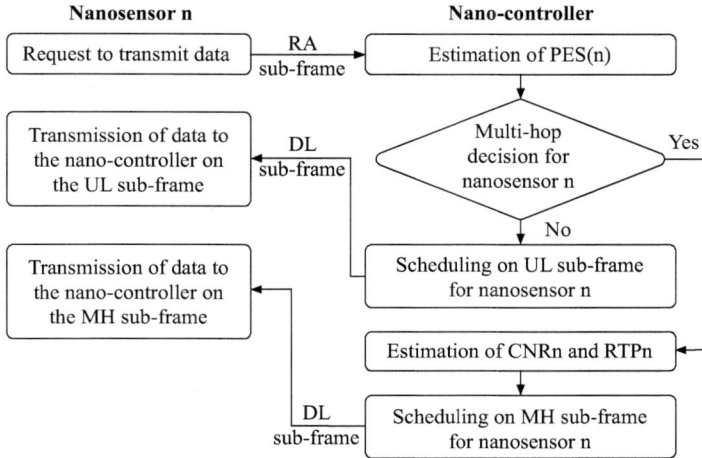

Figure 8.9   Block schematic of the multihop decision algorithm involving a nanonode n and the nanocontroller [6]

3. The nanocontroller decides for the nanosensor between multihop transmission and single-hop transmission. If the decision is single-hop transmission, the scheduling is performed on the UL subframe. If the decision is multihop transmission, the nanocontroller computes the critical neighborhood range of nanonode n (CNRn) to balance the throughput and infinite network lifetime. The required transmission power of the nanonode n (RTPn) is calculated according to the CNRn. Then, the nanocontroller sends the RTPn to nano-sensor n, and the nanosensor transmits the data in the MH subframe by using a transmission power equal to RTPn.

Then, nanosensor m is available to be the next hop only if all the following conditions are satisfied:

1. Nanosensor m has sufficient space in the retransmission buffer to store the data transmitted by nanosensor n.
2. Nanosensor m has sufficient energy to receive the data transmitted by nanosensor n.
3. The signal-to-noise ratio (SNR) of the channel between nanosensor m and nanosensor n (SNRn) is higher than a predefined received SNR value.
4. Nanosensor m is closer to the nanocontroller than is nanosensor n. The comparison of the distances from the nanonodes to the nanocontroller is based on the received signal strength indicator (RSSI), which is a measurement of the transmission power received at a nanosensor.

In this routing framework, the nanonodes adjust the transmission power and select the next hop node according to their available energy and current loads in multihop transmission. The simulation results show that this routing framework can reduce the average energy cost and increase the overall throughput [6]. However, the computational complexity of this routing framework is high, which may limit its practical application.

8.3.3.2   Energy-efficient multihop routing

Like in MHTD, a nanocontroller is assumed in an energy-efficient multihop routing (EEMR) protocol [34]. In the EEMR protocol, calculations are performed mainly in the nanocontrollers to reduce the computational complexity in the nanonodes. The difference from MHTD is that the distance between a nanonode and the nanocontroller is stored in the nanonode, and the one-hop range of the nanonode is fixed to reduce the amount of computation. According to the distance from the source node to the nanocontroller and the one-hop range of the source node, three regions exist, as shown in Figure 8.10. Region A1 is a circle whose radius is the distance from the nanocontroller to the source node. Regions A2 and A3 are within the one-hop range of the source node, but the nanonodes in A3 are closer to the nanocontroller than to the source node. In the EEMR protocol, the source node transmits data to a nanonode in A3, and the nanonode that receives the data becomes the next source node. The details of the data routing process are as follows:

1. In the initial phase, nanocontroller vc broadcasts hello messages and records the node ID and location that are sent back by the nanonodes.
2. When nanonode vi has data to send, vi first confirms whether vc is in its one-hop range. If so, vi directly sends data to vc. If not, vi broadcasts query messages to its neighboring nanonodes.
3. The neighboring nodes in region A3 are candidate nanonodes that may become the next source node. The candidate nanonodes calculate their link costs and send ACK messages, including the link cost and their locations, to vi.

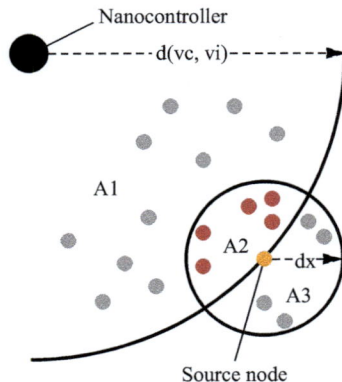

Figure 8.10   Area of candidate nodes [34]

4. After vi receives the ACK messages, its neighboring nanonodes are sorted in ascending order of link cost. Then, vi selects the first m candidate nanonodes with the smallest link costs and calculates the forwarding probabilities.
5. Nanonode vi randomly sends the data to one of the candidate nanonodes according to the forwarding probabilities; finally, the candidate nanonode that receives the data becomes the next source node.

The EEMR protocol limits the region of the next-hop candidate nanonode to control the direction of multihop forwarding. Although the energy efficiency of EEMR is not as high as that of MHTD in theory, the computational complexity of EEMR is much lower than that of MHTD.

8.3.3.3 Energy-conserving routing

A new application of nanonetworks is wireless body sensor nanonetworks (WBSNs) [36–38]. In WBSNs, terahertz waves transmitted through human tissues are attenuated. The path loss includes mainly molecular absorption attenuation, spreading loss and shadowing impact.

In WBSNs, nanosensors are assumed to be implanted in human body, and hierarchical architecture is used to maintain efficient nanosensor communication. An energy-conserving routing (ECR) protocol [38] was designed for WBSNs. In the ECR protocol, the WBSN consists of a nanocontroller (NC), several nanocluster controllers (NCCs) and nanocluster members (NCMs). Two types of communication occur in the ECR protocol, namely, intracluster communication and intercluster communication.

The ECR protocol uses a multilayer topology, as shown in Figure 8.11. The division of layers is based on the single-hop transmission range of the nanosensors and the distance from the nanosensors to the NC, and the width of each layer is one-half of the single-hop transmission range.

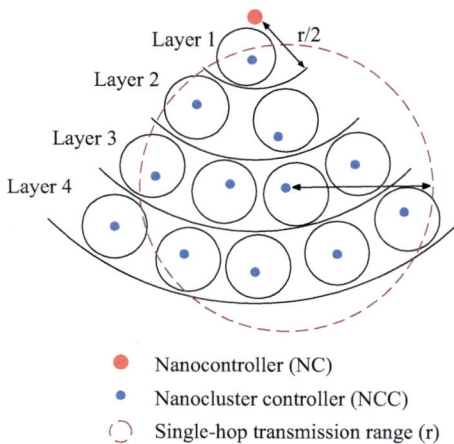

Figure 8.11   The architecture of ECR

After the division of layers, the nanosensors with remaining energy are selected to be the NCCs in the initial round. Then, the NCCs broadcast short-range advertisement messages. The received signal strength indicator (RSSI) is used to measure the signal strength of the advertisement messages from several NCCs. Each nanosensor selects the NCC with the highest RSSI. Then, the nanosensor sends a join request to the selected NCC to register itself as an NCM.

In intercluster communication, the NCC transmits the data to the NCCs in the lower layer. However, in the WBSNs, several nanosensors always have data to transmit in different clusters. In the ECR protocol, the NC sorts the transmission order of each layer according to the priority of the data and allocates a transmission time slot to each layer according to the total time required for data transmission. Then, in each layer, the transmission time slot is allocated for each cluster in the same way.

In intracluster communication, since the NCMs are close to the NCC in a cluster, only one-hop or double-hop transmissions are considered. When an NCM has data to transmit, a double-hop transmission may replace direct transmission if the double-hop transmission consumes less energy.

Since the cross-layer transmission and the collection of data in the cluster require more energy, the remaining energy of the NCCs is less than that of other nanosensors. Therefore, NCCs must be re-selected after completing each transmission.

The multi-layered topology and transmission time allocation of ECR effectively reduce the data transmission collisions and balance the energy consumption in WBSNs.

8.3.3.4   TEFoward

TEFoward: TTL-based efficient forwarding (TEFoward) [39] was designed for multihop polling-based EM-WNSNs [40] with a low density of nanosinks in the Internet of things (IoT). In the EM-WNSN architecture, nanosinks aggregate sensor data from nanosensors and forward the data to the IoT gateway [41]. In the TEFoward protocol, the latest topology information is reflected by polling the beaconing duplicate count and the TTL value of the nodes, which allows the nanosinks to select forwarders to transmit data packets. The aim of the TEForward protocol is to reduce the number of nanosinks that receive packets under dynamic channel states. During each polling interval, the nanosinks infer the latest network topology information by extracting the duplicate count and TTL values for forwarder selection and data diffusion.

As shown in Figure 8.12(a), in each polling process, the IoT gateway transmits a polling beacon for data extraction. The polling beacon has the packet format shown in Figure 8.13, and the TTL value of the beacon is set to TTLmax. In addition to the original fields of the packet header [42], a new field, called CumNP, is added to the polling beacon in TEForward. CumNP, which represents the total number of nanosinks that receive message packets during data delivery, plays an important role in forwarder selection. Furthermore, CumNP indicates the end-to-end energy consumption caused by the packet reception process and can be used to identify the forwarder that directs packets along the path that triggers the minimum

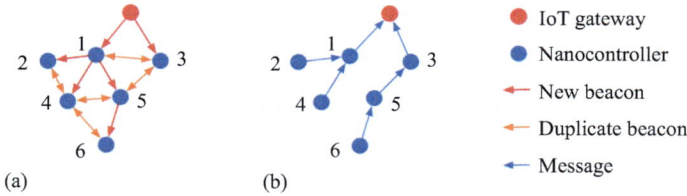

Figure 8.12 *Data dissemination of TEForward [39]. (a) Beacon dissemination and (b) message forwarding*

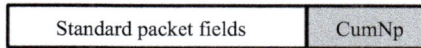

Figure 8.13 *Structure of a polling beacon [39]*

number of packet receptions. CumNF, which is the cumulative neighbor sizes of the nanosinks, is set to 0 by the IoT gateway and is updated by the nanosinks to cumulate the local neighbor sizes during beacon flooding. The nanosink that receives a new beacon performs the following steps:

1. First, the nanosink aligns its TTL settings TTLS to obtain the distance from the IoT gateway based on the TTL value of the beacon packet TTLP.
2. The nanosink resets the local neighbor size NS and initializes the local variable CumNF with CumNP in Beacon.

If a nanosink receives a duplicate of beacon, the sink performs the following steps:

1. The nanosink compares the value of TTLS with TTLP to check whether the beacon duplicate comes from a sink that is located close to the IoT gateway.
2. The nanosink compares the value of CumNF with CumNP to check if the sink that the beacon duplicate comes from provides a path with higher energy efficiency than the last forwarding candidate.
3. If so, the nanosink selects the beacon sender as its forwarder by recording the MAC source ID of the duplicate beacon.
4. The duplicate beacon is then dropped.

An example of forwarder selection is depicted in Figure 8.12(b), where "Nanosink 5" selects "Nano-sink 3" as its forwarder rather than "Nano-sink 1" because the cumulative neighbor size of "Nano-sink 3" is lower than that of "Nano-sink 1", which mitigates the number of nanosinks involved in packet reception during end-to-end data delivery.

The final selected forwarder directs packets toward the IoT gateway with high energy efficiency due to the minimized number of nanosinks activated for packet transmission and reception. The nanosinks that are involved in packet forwarding forward the message packet to their respective selected forwarders. In this way, all message packets are effectively and efficiently directed to the IoT gateway along the latest efficient path of the polling moment. With minimal computational resource support, the TEFoward protocol can ensure both the connectivity and efficiency of data transfers in dynamic channel states.

8.3.4 Comparison of routing protocols

The existing proposed routing protocols for WNNs are based mostly on the flooding scheme or single-path-based scheme, which aim at reducing energy consumption. The routing protocols include limit-flood-area-based protocols (e.g., RADAR, CORONA, SLR), DIF-based protocols (e.g., DEROUS, LSDD), and single-path-based protocols (e.g., MHTD, EEMR, ECR, TEForward). The routing framework proposed by Pierobon *et al.* [6] can help other researchers to develop new routing protocols.

To visually compare the existing proposed routing protocols, Table 8.1 summarizes all routing protocols reviewed in this chapter. The metrics for these routing protocols are described in the following:

- Forwarding strategy. In some routing protocols, nanonodes use a forwarding strategy to determine how to forward packets. In the DEROUS and LSDD protocols, the nanonodes that are self-classified as dynamic infrastructure retransmit the data. In the CORONA and SLR protocols, a nanonode that receives a packet for the first time compares its own coordinates with the coordinate value in the packet to determine whether to retransmit the packet. In the MHTD protocol, the transmission mode is first determined to be multihop or single-hop. Then, the data transmission path is based on the critical neighborhood range. In the ECR protocol, intracluster transmission is either single-hop or double-hop, depending on which consumes less energy. The intercluster transmission is layer-by-layer. In the EEMR protocol, the next hop in the transmission is the nanonode that is in the one-hop range of the source node and close to the nanocontroller. In the TEForward protocol, the data transmission path is based on the cumulative neighbor size.
- Position awareness. A nanonetwork with nanonodes that have position information can control the transmission better, for example, by limiting the flooding area. In the DEROUS, CORONA, and SLR protocols, nanonodes are aware of their hop counts to the central entity or anchor nodes. In the ECR protocol, nanonodes are aware of which layer they are in. In the EEMR protocol, nanonodes are aware of their distance to the nanocontroller.
- Load balance. When a routing protocol contains optimization for network load balancing, all nanonodes in the nanonetwork can maintain the balance of the remaining energy while not allowing excessive energy consumption by some nanonodes. In the DEROUS and LSDD protocols, only the nanonodes with high residual energy can become part of the dynamic infrastructure and transmit data; thus, load balancing can be achieved. In the hierarchical routing protocols, load balancing is achieved by the rotation of the cluster head and multihop transmission decision.

Table 8.1 Comparison of each protocol

Reference	Routing protocol	Routing strategy	Forwarding strategy	Position awareness	Network structure	Load balance	Computational complexity	Scalability
[43]	RADAR routing	Limit-flood-area-based	Blindly retransmit	No	Flat	No	Low	Limited
[23]	CORONA	Limit-flood-area-based	Coordinate	Hop counts to the anchor nodes (coordinate)	Flat	No	Low	Limited
[28,29]	SLR	Limit-flood-area-based	Coordinate	Hop counts to the anchor nodes (coordinate)	Flat	No	Medium	Limited
[31]	LSDD	DIF-based	Dynamic infrastructure	No	Flat	Yes	Low	Good
[34]	DEROUS	DIF-based	Dynamic infrastructure	Hop counts to the central entity (radius)	Flat	Yes	Low	Good
[6]	MHTD	Single-path-based	Multihop decision algorithm	No	Hierarchical	Yes	High	Good
[36]	EEMR	Single-path-based	Direction of forwarding	Distance to the nanocontroller	Hierarchical	No	High	Good
[37,38]	ECR	Single-path-based	Double-hop or one-hop (intracluster) Layer-by-layer (intercluster)	Layer	Hierarchical	Yes	Medium	Good
[40]	TEForward	Single-path-based	Cumulative neighbor size	No	Hierarchical	No	Medium	Good

- Computational complexity. The effectiveness of the protocol designed under the required constraints is determined by the complexity of the protocol. However, nanonodes in the nanonetwork have fewer computing resources, power, and memory capacity and therefore require a simple routing protocol. The ratings are based on the following criteria. In the forwarding policy, if the nanonode do not need to perform calculations or only needs to make a simple comparison, the computational complexity of routing protocol is "Low." If the nanonode only needs simple calculations, the computational complexity of routing protocol is "Medium." If the nanonode needs other nodes to assist in the calculation, the computational complexity of routing protocol is "High." RADAR routing is a simple flood-based routing protocol. In the LSDD and DEROUS protocols, nanonodes are self-classified as "infrastructure" or "user" based on their packet-receive statistics. In the CORONA protocol, nanonodes only need to compare their coordinates with the coordinates in the data packet. In the SLR, ECR, EEMR, and TEForward protocols, nanonodes must perform simple calculations, comparisons, and judgments. In the MHTD protocol, the nanocontroller calculates the probability of saving energy through multihop transmission, the critical neighborhood range, the random back-off time for node load fairness, etc. Therefore, RADAR routing and the DEROUS, LSDD, and CORONA protocols have low complexity; the SLR, ECR, EEMR, and TEForward protocols have medium complexity; and the MHTD protocol has high complexity.
- Scalability. Scalability refers to the ability of the protocol to accommodate increasing workloads, which means that the protocols are expected to achieve good performance in both small and large networks. The ratings are based on the following criteria. As the number of nanonodes increases, the scalability of routing protocol is "Limited" if the network performance drops significantly. If the network performance drops slightly, the scalability of the routing protocol is "Good." In RADAR routing and the CORONA and SLR protocols, when the nanonetwork is large, the flooding area is also large. In the DEROUS and LSDD protocols, despite the flood-based transmission, only nodes with high residual energy retransmit data. The hierarchical routing protocols always have good scalability owing to the hierarchical nanonetwork structure. Therefore, the DEROUS, LSDD, MHTD, ECR, EEMR, and TEForward protocols have good scalability, and RADAR routing and the CORONA and SLR protocols have limited scalability.

8.4 Conclusion

In this chapter, we focus on the routing protocols that can be applied in WNNs: RADAR routing, CORONA, SLR, LSDD, DEROUS, MHTD, EEMR, ECR, and TEForward. Each of the protocols has advantages and disadvantages.

In WNNs, the small size of the nanonodes leads to problems such as limited computing resources and low energy storage. To guarantee the connection of the nanonodes, new routing protocols with low complexity and energy consumption must be designed.

Therefore, further study on designing efficient routing protocols in WNNs and improving the energy efficiency of nanonodes is essential. In WNN routing protocols, "hop count" is a parameter that can be well utilized. The "hop count" can be used to design routing protocols that limit the flooding area, and in combination with the DIF method, the routing protocols can even control the data transmission direction. In future research, the "hop count" and DIF can be used to design WNN routing protocols. In single-path routing, RSSI can be used to transfer data toward the nanocontroller. In addition, it is necessary to consider the remaining energy of nanonodes, the hop counts to the nanocontroller, and the power used for each transmission to find a suitable transmission path.

Although several routing protocols for WNNs have been proposed, many open questions remain to be studied. The proposed routing protocols are all designed to reduce energy consumption, but other aspects have not been considered. Some other aspects, such as mobile nanonodes and quality of service, may deserve to be studied. We provide several suggestions for future research on routing protocols in the following.

Since routing protocols are designed based on the network architectural model, the performance of the routing protocols is closely related to the network architecture. The network architecture is designed depending on the application of the nanonetwork. For different applications, some important factors that affect the selection of the routing protocol include the following:

- Nanonode deployment: Nanonode deployment may be deterministic or random, which may impact the performance of the routing protocol. In deterministic situations, routing paths can be predefined by placing the nodes manually. In random node deployment, which is the deployment method used in most nanonetworks, the nanonodes form the nanonetwork in an ad hoc manner.
- Nanonode capability: A nanonode usually has a specific function in a WNN, such as sensing, relaying, and aggregation, due to the restrictions of energy and computing resources.
- Network dynamics: Nanonetworks have three prime components: nanonodes, nanosinks, and monitored events. The nanonodes are assumed to be stationary in most WNNs. In mobile-nanonode nanonetworks, the mobility of nanonodes may be intentional or unintentional. On the other hand, the sink nodes or cluster heads are sometimes assumed to be mobile. Routing a message among nanonodes is challenging due to the need to optimize the routing stability. Finally, the monitored event's dynamics are generally determined by application because the required monitoring data may be dynamic or static.
- Energy considerations: Energy considerations greatly influence the selection of a routing protocol. The communication distance and the presence of obstacles are important factors for the transmission power of wireless radio. In contrast to multihop routing, which consumes less energy, direct routing performs well when all the nanonodes are close to the nanosink. In addition, nanonodes

have a special energy harvesting processes and low energy storage. These two characteristics must be considered when developing new routing protocols.

- Data delivery model: Four data delivery models exist—the continuous, event-driven, query-driven, and hybrid models. Nanosensors in the continuous model send data periodically. In the event-driven and query-driven models, the transmission of data is influenced by an event or query generated by the nanosinks. The hybrid model is a combination of continuous, event-driven, and query-driven models that is applied in some nanonetworks. Under these conditions, the minimization of energy consumption and routing stability are essential and necessary for the routing protocol.

The general goal of routing protocols is to transmit information among nodes in a variety of ways. However, due to the limitations of WNNs, no routing table exists; therefore, the topology information cannot be determined by nanonodes. To overcome the limitations of WNNs, the goals for developing new routing protocols are listed as follows:

- reduce energy consumption and consider load balancing to increase the lifetime of the nanonetwork;
- reduce protocol complexity and improve robustness;
- minimize the transmission delay.

In addition to the above goals, many problems are worth additional research, not only in terms of energy efficiency. During the development of a new routing protocol, the following points should be considered, checked, and examined.

- Energy consumption balance: The most important and essential target during the process of data routing is how to balance the energy consumption of the nanonodes. To ensure an energy consumption balance, new routing protocols should attempt to use nanonodes with high residual energy for data routing.
- Node mobility: We always assume that the nodes are static. However, researchers have increasingly focused on applications with mobile nodes in WSNs, e.g., sink mobility has been recognized as an efficient method of improving system performance in WSNs [44]. Therefore, new protocols for WNNs also need to be designed and applied in these conditions.
- Protocol complexity: Due to the limited computing resources of nanonodes, the complexity of the routing protocol cannot be too high, which must be considered when designing new routing protocols.
- Performance evaluation in a real environment: Although some routing protocols can achieve satisfactory performance in simulations, applying the new routing protocols in a real environment remains challenging.
- Quality of service: The definition of QoS is slightly different in various application areas [45]. In routing protocols, QoS is an important metric of the performance, which means that energy efficiency and accurate delivery should be considered when designing new routing protocols.

References

[1]   C. R. Yonzon, D. A. Stuart, X. Zhang, A. D. Mcfarland, C. L. Haynes, and R. P. Van Duyne, "Towards advanced chemical and biological nanosensors-an overview," Talanta, vol. 67, no. 3, pp. 438–448, 2005.

[2]   J. Riu, A. Maroto, and F. X. Rius, "Nanosensors in environ-mental analysis," Talanta, vol. 69, no. 2, pp. 288–301, 2006.

[3]   C. Hierold, A. Jungen, C. Stampfer, and T. Helbling, "Nano electromechanical sensors based on carbon nanotubes," Sensors and Actuators A Physical, vol. 136, no. 1, pp. 51–61, 2007.

[4]   C. Li, E. T. Thostenson, and T. W. Chou, "Sensors and actuators based on carbon nanotubes and their composites: A review," Composites Science and Technology, vol. 68, no. 6, pp. 1227–1249, 2008.

[5]   I. F. Akyildiz and J. M. Jornet, "Electromagnetic wireless nanosensor networks," Nano Communication Networks, vol. 1, no. 1, pp. 3–19, 2010.

[6]   M. Pierobon, J. M. Jornet, N. Akkari, S. Almasri, and I. F. Akyildiz, "A routing framework for energy harvesting wireless nanosensor networks in the terahertz band," Wireless Networks, vol. 20, no. 5, pp. 1169–1183, 2014.

[7]   P. Suriyachai, U. Roedig, and A. Scott, "A survey of mac protocols for mission-critical applications in wireless sensor net-works," IEEE Communications Surveys and Tutorials, vol. 14, no. 2, pp. 240–264, 2012.

[8]   F. Dressler and F. Kargl, "Towards security in nano-communication: Challenges and opportunities," Nano Communication Networks, vol. 3, no. 3, pp. 151–160, 2012.

[9]   C. Rutherglen and P. Burke, "Nanoelectromagnetics: Circuit and electro-magnetic properties of carbon nanotubes," Small, vol. 5, no. 8, pp. 884–906, 2009.

[10]   I. F. Akyildiz, F. Brunetti, and C. Blázquez, "Nanonetworks: A new com-munication paradigm," Computer Networks, vol. 52, no. 12, pp. 2260–2279, 2008.

[11]   C. N. R. Rao and A. Govindaraj, "Nanotubes and nanowires," Journal of Chemical Sciences, vol. 59, no. 22, pp. 4665–4671, 2001.

[12]   J. M. Jornet and I. F. Akyildiz, "Graphene-based nano-antennas for electro-magnetic nanocommunications in the terahertz band," in European Conference on Antennas and Propagation, 2010, pp. 1–5.

[13]   J. M. Jornet and I. F. Akyildiz, "Channel modeling and capacity analysis for electro-magnetic wireless nanonetworks in the terahertz band," IEEE Transactions on Wireless Communications, vol. 10, no. 10, pp. 3211–3221, 2011.

[14]   P. Boronin, V. Petrov, D. Moltchanov, Y. Koucheryavy, and J. M. Jornet, "Capacity and throughput analysis of nanoscale machine communication through transparency windows in the terahertz band," Nano Communication Networks, vol. 5, no. 3, pp. 72–82, 2014.

[15] X.-W. Yao, C.-C. Wang, W.-L. Wang, and C. Han, "Stochastic geometry analysis of interference and coverage in terahertz networks," Nano Communication Networks, vol. 13, pp. 9–19, 2017.

[16] I. F. Akyildiz, J. M. Jornet, and C. Han, "Terahertz band: Next frontier for wireless communications," Physical Communication, vol. 12, no. 4, pp. 16–32, 2014.

[17] J. M. Jornet and I. F. Akyildiz, "Joint energy harvesting and communication analysis for perpetual wireless nanosensor networks in the terahertz band," IEEE Transactions on Nanotechnology, vol. 11, no. 3, pp. 570–580, 2012.

[18] M. Lallart, and D. Guyomar, "Nonlinear energy harvesting," IOP Conference Series: Materials Science and Engineering, vol. 18, no. 9, 092006, 2011.

[19] L. Gammaitoni, I. Neri, and H. Vocca, "Nonlinear oscillators for vibration energy harvesting," Applied Physics Letters, vol. 94, no. 16, 164102, 2009.

[20] Z. L. Wang, "Towards self-powered nanosystems: From nano-generators to nanopiezotronics," Advanced Functional Materials, vol. 18, no. 22, pp. 3553–3567, 2008.

[21] X.-W. Yao, W.-L. Wang, and S.-H. Yang, "Joint parameter optimization for perpetual nanonetworks and maximum network capacity," IEEE Transactions on Molecular, Biological and Multi-Scale Communications, vol. 1, no. 4, pp. 321–330, 2015.

[22] L.-J. Huang, X.-W. Yao, W.-L. Wang *et al.*, "Eoc: Energy optimization coding for wireless nanosensor networks in the terahertz band," IEEE Access, vol. 5, pp. 2583–2590, 2017.

[23] A. Tsioliaridou, C. Liaskos, S. Ioannidis, and A. Pitsillides, "Corona: A coordinate and routing system for nanonetworks," in International Conference on Nanoscale Computing and Communication, 2015, pp. 1–6.

[24] X.-W. Yao, C.-C. Wang, W.-L. Wang, and J. M. Jornet, "On the achievable throughput of energy-harvesting nanonetworks in the terahertz band," IEEE Sensors Journal, vol. 18, no. 2, pp. 902–912, 2018.

[25] N. A. Pantazis and D. D. Vergados, "A survey on power control issues in wireless sensor networks," IEEE Communications Surveys and Tutorials, vol. 9, no. 4, pp. 86–107, 2007.

[26] N. A. Pantazis, S. A. Nikolidakis, and D. D. Vergados, "Energy-efficient routing protocols in wireless sensor networks: A survey," IEEE Communications Surveys and Tutorials, vol. 15, no. 2, pp. 551–591, 2013.

[27] S. R. Neupane, "Routing in resource constrained sensor nanonetworks," Applied Soft Computing, vol. 26, pp. 285–298, 2014.

[28] A. Tsioliaridou, C. Liaskos, E. Dedu, and S. Ioannidis, "Stateless linear-path routing for 3d nanonetworks," in ACM International Conference on Nanoscale Computing and Communication, 2016, pp. 1–6.

[29] A. Tsioliaridou, C. Liaskos, E. Dedu, and S. Ioannidis, "Packet routing in 3d nanonetworks: A lightweight, linear-path scheme," Nano Communication Networks, vol. 12, pp. 63–71, 2017.

[30] S. Crisóstomo, U. Schilcher, C. Bettstetter, and J. Barros, "Probabilistic flooding in stochastic networks: Analysis of global information outreach," Computer Networks the International Journal of Computer and Telecommunications Networking, vol. 56, no. 1, pp. 142–156, 2012.

[31] A. Tsioliaridou, C. Liaskos, S. Ioannidis, and A. Pitsillides, "Lightweight, self-tuning data dissemination for dense nanonet-works," Nano Communication Networks, vol. 8, pp. 2–15, 2016.

[32] J. Misra and D. Gries, "Finding repeated elements," Science of Computer Programming, vol. 2, no. 2, pp. 143–152, 1982.

[33] C. Liaskos and A. Tsioliaridou, "A promise of realizable, ultra-scalable communications at nano-scale: A multi-modal nano-machine architecture," Computers IEEE Transactions on Computers, vol. 64, no. 5, pp. 1282–1295, 2015.

[34] J. Xu, R. Zhang, and Z. Wang, "An energy efficient multi-hop routing protocol for terahertz wireless nanosensor networks," in International Conference on Wireless Algorithms, Systems, and Applications, 2016, pp. 367–376.

[35] C. Liaskos, A. Tsioliaridou, S. Ioannidis, N. Kantartzis, and A. Pitsillides, "A deployable routing system for nanonetworks," in 2016 IEEE International Conference on Communications (ICC), 2016, pp. 1–6.

[36] F. Afsana, N. Jahan, and M. S. Kaiser, "An energy efficient cluster based forwarding scheme for body area network using nano-scale electromagnetic communication," in 2015 IEEE International WIE Conference on Electrical and Computer Engineering (WIECON-ECE), 2015.

[37] G. Piro, G. Boggia, and L. A. Grieco, "On the design of an energy-harvesting protocol stack for body area nano-networks," Nano Communication Networks, vol. 6, no. 2, pp. 74–84, 2015.

[38] F. Afsana, M. Asifurrahman, M. R. Ahmed, M. Mahmud, and M. S. Kaiser, "An energy conserving routing scheme for wireless body sensor nanonetwork communication," IEEE Access, vol. 6, pp. 9186–9200, 2018.

[39] H. Yu, B. Ng, W. K. G. Seah, and Y. Qu, "TTL-based efficient forwarding for the backhaul tier in nanonetworks," in 2017 14th IEEE Annual Consumer Communications and Networking Conference (CCNC), 2017.

[40] C. Fujii and W. K. G. Seah, "Multi-tier probabilistic polling in wireless sensor networks powered by energy harvesting," in 2011 Seventh International Conference on Intelligent Sensors, Sensor Networks and Information Processing, 2011.

[41] I. F. Akyildiz and J. M. Jornet, "The internet of nano-things," IEEE Wireless Communications, vol. 17, no. 6, pp. 58–63, 2010.

[42] G. Piro, L. A. Grieco, G. Boggia, and P. Camarda, "Nano-sim: simulating electromagnetic-based nanonetworks in the network simulator 3," in International ICST Conference on Simulation TOOLS and Techniques, 2013, pp. 203–210.

[43] A. Tsioliaridou, C. Liaskos, L. Pachis, S. Ioannidis, and A. Pitsillides, "N3: Addressing and routing in 3d nanonetworks," in International Conference on Telecommunications, 2016, pp. 1–6.

[44]   Y. Gu, F. Ren, Y. Ji, and J. Li, "The evolution of sink mobility management in wireless sensor networks: A survey," IEEE Communications Surveys and Tutorials, vol. 18, no. 1, pp. 507–524, 2016.

[45]   I. Al-Anbagi, M. Erol-Kantarci, and H. T. Mouftah, "A survey on cross-layer quality-of-service approaches in WSNs for delay and reliability-aware applications," IEEE Communications Surveys and Tutorials, vol. 18, no. 1, pp. 525–552, 2016.

Chapter 9

Error-control mechanisms for nano-electromagnetic communication networks

Xin-Wei Yao[1], De-Bao Ma[1] and Chong Han[2]

Nanonetworks consist of nano-sized communication devices that perform simple tasks such as computation, data storage, and actuation at the nanoscale. However, communication in nanonetworks is constrained by error-prone wireless links due to severe path loss in the terahertz band (0.1–10.0 THz) and the very limited energy storage capacity of nanodevices. Therefore, efficient and effective error-control protocols are required for nanonetworks in the THz band. In this chapter, first, the related works on error control for nanonetworks are presented and investigated by considering the corresponding features. Second, a new error-control strategy with probing (ECP) mechanism for nanonetworks powered by energy harvesting is proposed. In particular, each data packet will not be transmitted until the communication of one probing packet is successful. Third, an energy state model is presented by considering the energy-harvesting–consumption process based on the extended Markov chain approach. Moreover, a probabilistic analysis of overall network traffic and multiuser interference is used by the proposed energy state model to capture dynamic network behavior. Following that, the impact of the energy consumption of different packets on state transition and the state probability distribution of nanonodes based on the above model are comprehensively investigated. Finally, the performance of the ECP mechanism is investigated and evaluated in terms of end-to-end successful packet delivery probability, end-to-end packet delay, achievable throughput, and energy consumption by comparing with other four different error-control strategies, such as automatic repeat request (ARQ), forward error correction (FEC), error prevention code (EPC), and a hybrid EPC (HEPC).

9.1 Introduction

Nanotechnology enables the development of nanonetworks [1,2], which contributes to its potential in many unprecedented applications. For example, there are many direct applications of nanotechnology in the biomedical field (e.g., drug delivery

[1]College of Computer Science and Technology, Zhejiang University of Technology, Hangzhou, China
[2]University of Michigan–Shanghai Jiao Tong University Joint Institute, Shanghai Jiao Tong University, Shanghai, China

systems and cancer detection), environmental field (e.g., plant monitoring systems and biodegradation), and military field (e.g., biological and chemical defenses and nano-functionalized equipment) [1,3]. To realize these applications and guarantee their performance, researchers should comprehensively consider the energy consumption, data processing, and communication issues in the nanodevices.

Recently, the THz band has been recommended as the communication band for nanodevices due to the recent development in the area of graphene-based nano-electronics [4,5]. However, it undergoes a severe propagation loss due to the atmospheric absorption [6], which results in a very limited communication distance [7]. According to the latest research on graphene-based nano-transceivers [8,9] and nano-antennas [5,10], the THz band has shown its great potential as the communication frequency band for nanodevices.

Consequently, the limited capabilities of a single nanodevice and the THz-band channel behavior easily lead to error-prone wireless links in nanonetworks. It is necessary to design a suitable error-control mechanism for nanonetworks. The existing error-control mechanisms of traditional wireless networks can be classified into three main approaches, that is, ARQ, FEC, and hybrid automatic repeat request (HARQ) [11], which are described as follows [12–14]:

- Automatic repeat request (ARQ): In ARQ-based error-control schemes, retransmission has a major impact on recovering lost packets. ARQ schemes implement retransmissions of failed packets by sending explicit acknowledgments. Obviously, ARQ schemes generate a lot of extra retransmission cost in case of errors. In addition to the overhead and delay caused by retransmissions, harvesting energy for retransmissions is another challenging task. In contrast, the overhead of ARQ schemes is lower than FEC protocols when the channel condition is good, because it does not have frequent retransmissions and has no need to add redundancy to the transmitted packets.

- Forward error correction (FEC): When using FEC-based error-control techniques, some redundant bits are added to the transmitted packets so that the receiver can recover the original bits even if a finite number of bits are received erroneously. Compared with ARQ-based techniques, FEC-based techniques require more complex computational capability and sacrifice data rates for coding appendant although they increase the communication distance and improve the communication error recovery.

- Hybrid automatic repeat request (HARQ): HARQ error-control techniques combine the advantages of ARQ and FEC techniques to increase the error resiliency while reducing the number of retransmissions. There are mainly two types of HARQ techniques. For the HARQ-I (called Chase combining) technique, the transmitter first sends a packet coded with a low error correction capability to the receiver. In the case of the reception of an NAK signal, the transmitter resends the same coded packet again. In the HARQ-II (referred to the incremental redundancy) technique, the transmitter constructs and resends different packets coded with a more powerful FEC code when an NAK is received. Generally, in HARQ-I and HARQ-II, failed packets are first stored in the buffer, and then, a retransmission is requested. In contrast, HARQ-I can improve link performance through combining multiple copies of the received packet, and HARQ-II can jointly utilize the differently coded versions of retransmitted packets to form a lower-rate code with stronger error protection capabilities.

As mentioned above, traditional ARQ, FEC, and HARQ techniques cannot be directly applied in nanonetworks due to the very limited computational capability and energy storage capacity of nanodevices. However, our proposed ECP mechanism is based on the traditional ARQ protocol, which comprehensively considers the peculiarities of nanonetworks in energy and computation.

In this chapter, a novel error-control mechanism is proposed by considering the trade-off between energy harvesting and consumption for perpetual nanonetworks. The main contributions of ECP are summarized as follows:

1. A novel error-control mechanism with probing for nanonetworks powered by energy harvesting is proposed. In particular, before starting data transmission, one probing packet is sent out to detect the channel status. Consequently, the corresponding feedback of the probing packet decides whether the data packet will be transmitted.
2. Energy state model of the ECP mechanism by considering the energy-harvesting–consumption process is presented based on the extended Markov chain model. Moreover, the proposed model captures the dynamic network behavior through a probabilistic analysis of the total network traffic and the multiuser interference in the THz nanonetworks. The impact of the energy consumption of different packets on state transition and the probability distribution of every transmission state are investigated.
3. The energy state model is validated by means of simulation, and we numerically study the data packet size for the ECP mechanism and other four error-control techniques (ARQ, FEC, EPC, and HEPC) and compare their performance under different parameters. Performance analysis shows that the ECP mechanism outperforms traditional error-control schemes such as ARQ and FEC in terms of end-to-end successful packet delivery probability and energy savings while increases the end-to-end packet delay.

9.2   Related work

Despite the existence of many studies on error-control techniques for traditional wireless networks, none of them can be directly applicable to the nanonetworks because of the very limited capabilities of nanodevices and the channel behavior of the THz band, which are easy to lead to error-prone wireless links. The objective of this chapter is to propose a novel error-control mechanism that comprehensively considers the energy-harvesting and consumption process of nanonodes. Without loss of generality, the main relevant works are discussed next.

9.2.1   Existing work on error control in nanonetworks

Recently, there has been some work on error control for nanonetworks. In detail, a new error-control strategy for electromagnetic nanonetworks is proposed in [15].

This strategy is intended to prevent channel errors by utilizing low-weight channel codes. The results of this work reveal that the codeword error rate is reduced through the utilization of low-weight channel codes while the achievable information rate does not decrease or even increase it. Furthermore, an optimal code weight is found, which can maximize the information rate. In [16], the authors explored the optimal coding design for transmission energy minimization (MTE) in nanonetworks and developed the corresponding solutions by considering code rate constraint and codeword length constraint. The simulation results show that MTE coding can decrease the transmission energy consumption in nanonetworks while guaranteeing an acceptable code rate. Moreover, a cross-layer analysis of error-control strategies for nanonetworks in the THz band is presented in [12]. Alternatively, in [17], the authors addressed the link throughput maximization problem by considering the peculiarities of nanosensors and the optimal data packet size, which maximizes the link efficiency is investigated.

However, for the above existing work on error control, on the one hand, they either do not consider the energy-harvesting–consumption process or have no cognition on the temporary feature of energy, i.e., the process of energy-harvesting–consumption is dynamic. On the other hand, the channel condition is unknown before the data packet is transmitted, and the impact of transmission of different packets on the energy status of nanonode has not been investigated. Note that, in the simulations and performance evaluation, we use the traditional ARQ and FEC schemes as well as the new EPC schemes with constant low-weight channel codes [15] and an HEPC, which combines both FEC and EPC [12].

9.2.2   *Energy harvesting with piezoelectric nanonetworks*

It is highly desirable that nanodevices are enabled to be self-powered without the use of nanobatteries. However, traditional energy-harvesting mechanisms, e.g., solar energy, wind power, or underwater turbulence [18,19], cannot be utilized in nanonetworks. In recent years, some methods of converting mechanical energy into electrical energy have been explored [20,21], such as zinc oxide nanogenerators using the piezoelectric effect. The energy harvested from the external environment is usually sufficient to power the communication of nanodevices, especially over a short distance.

According to the prototype design and the corresponding circuit model of piezoelectric nanogenerators in [22,23], the harvested energy can be stored in the nanocapacitor of nanodevices. In general, the maximum energy E_{max} can be calculated as a function of the total capacitance C_{cap} and the generator voltage V_g [24], i.e.,

$$E_{max} = \max\left\{\frac{1}{2}C_{cap}\left(V_{cap}\left(n_{cyc}\right)\right)^2\right\} = \frac{1}{2}C_{cap}V_g^2 \qquad (9.1)$$

Finally, the energy-harvesting rate at the nanocapacitor can be given in Joule/second as follows:

$$\lambda_{harv}(E_{curr}, \Delta E) = \left(\frac{n_{cyc}}{t_{cyc}}\right)\frac{\Delta E}{n_{cyc}(E_{curr} + \Delta E) - n_{cyc}(E_{curr})} \qquad (9.2)$$

where E_{curr} is the current energy in the nanocapacitor, ΔE refers to the energy increment of the capacitor, and t_{cyc} is the time between consecutive cycles. n_{cyc} is the number of cycles that needed to charge the nanocapacitor up to an energy value E_{curr}. Our starting point for energy harvesting in our analysis is the energy model introduced in [24], which can clearly reproduce the energy-harvesting–consumption process.

9.2.3 *Energy consumption in pulse-based nanonetwork communication*

Due to the constrained energy in nanodevices, short pulses cannot be emitted in a burst. Hence, new information encoding and modulation mechanisms for nano-networks are imperatively required. Furthermore, in order to take advantage of the large bandwidth of THz band, a pulse-based communication paradigm (such as TS-OOK) has been recommended in [25], which is based on the exchange of very short pulses spread in time. In detail, a logical "1" is transmitted by a short pulse (100 fs), and a logical "0" is transmitted as silence, i.e., nanodevices remain silent when a logical "0" is transmitted.

In order to evaluate the consumed energy in the transmission and reception of one packet, without loss of generality, it is assumed that the length of one packet is N_{bits}^{y}, and the consumed energy in the transmission and reception of one pulse are E_{pul-t} and E_{pul-r}, respectively. Consequently, the consumed energy for transmitting and receiving one packet can be given by

$$\begin{cases} E_{tx}^{y} = N_{bits}^{y} W_{y} E_{pul-t} \\ E_{rx}^{y} = N_{bits}^{y} E_{pul-r} \end{cases} \tag{9.3}$$

where $y = p$ or $y = d$ stands for the calculation of one probing packet and one data packet, respectively. W_{y} refers to the coding weight, which is adaptive to different coding algorithms or parameter optimizations [26].

9.3 Error control with probing

In this section, a detailed description of the proposed error-control mechanism and its corresponding energy state model are presented. Through the energy state model, the trade-off between energy harvesting and energy consumption can be investigated comprehensively. Alternatively, the overall network performance is captured successfully by considering the total network traffic and the multiuser interference. Specifically, a mathematical framework is developed and used for the steady-state analysis of the model.

9.3.1 *Error-control mechanism*

In this section, the energy state model of the proposed ECP mechanism based on the extended Markov chain approach is presented by considering the energy-harvesting–consumption process. The associated notations in Figure 9.1 are listed

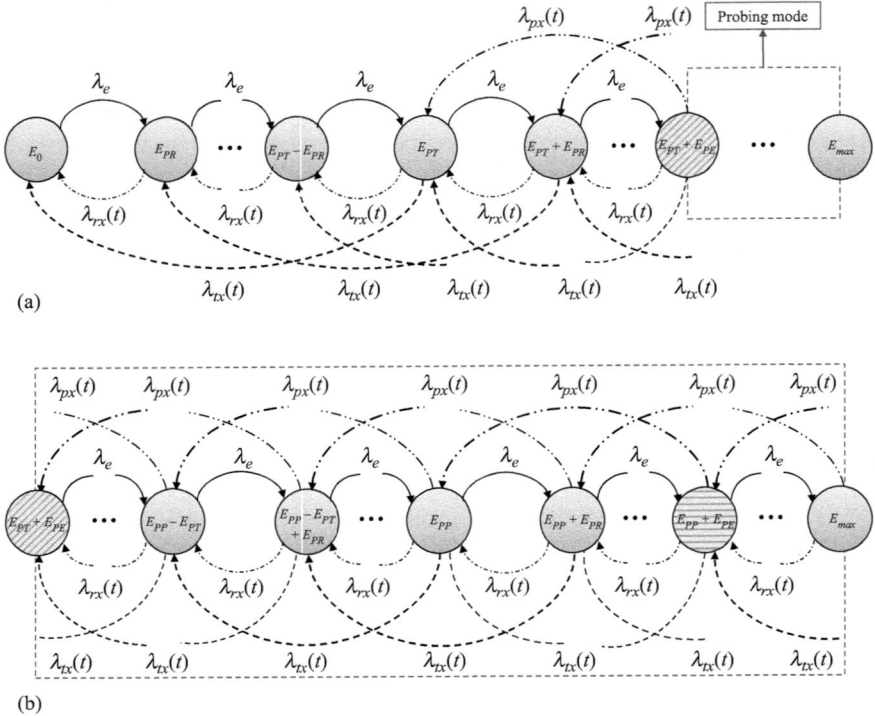

(a)

(b)

*Figure 9.1   (a) The general energy state model in the ECP mechanism. (b) Energy
state model of probing mode in ECP mechanism*

in Table 9.1. Each state in the model corresponds to an energy state of nanonode. The dashed red box in Figure 9.1(a) refers to a set of states that the nanonode works in the probing mode, which is shown comprehensively in Figure 9.1(b). The solid lines represent the energy-harvesting process, whereas the other lines represent the energy consumption process. The short dashed lines represent the energy consumption processes of sending one data packet, which is denoted as λ_{tx}. The dashed-dotted lines represent the energy consumption processes of receiving one data packet or one probing packet (in this chapter, it is assumed that the energy consumption of receiving one data packet and one probing packet is identical), which is denoted as λ_{rx}. The double dotted, dashed lines represent the energy consumption processes of sending one probing packet, which is denoted as λ_{px}. The detailed operations of the proposed ECP for electromagnetic nanonetworks are conducted as follows.

In order to increase the successful transmission probability of data packets and reduce the energy consumption of communication, the proposed ECP divides all energy states into two parts as shown in Figure 9.1. On the one hand, each

Table 9.1 Notations of the symbols in Figure 9.1

Symbol	Description
$\{E\}$	The energy state when its energy value is E
E_0	Energy value of initial state
E_{PR}	Energy consumption of receiving one packet, which equals to the data receiving energy threshold τ
E_{PT}	Energy consumption of sending one packet, which equals to the data packet sending energy threshold γ
E_{PE}	Energy consumption of sending one probing packet
E_{max}	Energy saturation value
$E_{PT} + E_{PE}$	Energy threshold of the probing mode, δ
E_{PP}	An energy value between $E_{PT} + E_{PE}$ and E_{max}
λ_e	Packet energy-harvesting rate
$\lambda_{tx}(t)$	The transmission rate of one data packet
$\lambda_{rx}(t)$	The reception rate of one data packet
$\lambda_{px}(t)$	The transmission rate of one probing packet

nanonode will keep harvesting energy and enter the probing mode only as its energy value is beyond the probing mode energy threshold δ, which is defined in Table 9.1. On the other hand, in the probing mode, one probing packet will be transmitted to detect the channel condition at first. The data packet will be sent out only after the probing packet is transmitted successfully.

In general, two conditions in the energy state model are considered. The first condition is that the nanonode only keeps harvesting energy without transmitting or receiving any packet from the initial energy state $\{E_0\}$ to the saturated energy state $\{E_{max}\}$. The second condition is that the nanonode needs to transmit or receive data packets while harvesting energy. More specifically, there are two modes in the second condition as follows:

- Mode 1 (general mode): From the beginning of the initial energy state $\{E_0\}$, the nanonode captures energy and receives packets (when necessary) rather than transmits any packet until it reaches the maximum energy E_{max} and enters the saturated energy state $\{E_{max}\}$. Moreover, the nanonode will not receive packets until the energy is beyond the data receiving energy threshold τ.
- Mode 2 (probing mode): While harvesting energy, the nanonode needs to transmit data packets as well. If nanonode wants to transmit data in a certain state such as $\{E_{PP} + E_{PE}\}$. First, its stored energy has to reach the energy threshold δ of probing mode. After sending out one probing packet, the nanonode consumes the energy E_{PE} and enters into the energy state $\{E_{PP}\}$. According to the received feedback from the receiver after the probing packet transmitted, there are two cases:
  1. Case 1: If the feedback of the probing packet is a negative acknowledgment (NAK) or timeout, i.e., the channel is in the bad condition, and the nanonode needs to retransmit the probing packet. It is worth noting that

the nanonode has to determine whether its current energy reaches the energy threshold δ before detecting the channel using the probing packet. As the energy is sufficient for channel detection, the probing packet will be sent out. The probing process continues until an ACK is received or its maximum number of retransmission is reached. Otherwise, the nanonode needs to harvest energy until the energy threshold δ, and then the channel detection is performed again.

2. Case 2: If the feedback of the probing packet is an active acknowledgment (ACK) and the energy of nanonode is beyond the energy threshold γ, then one data packet will be transmitted. This process will consume the energy E_{PT}, and let the nanonode move to the state $\{E_{PP} - E_{PT}\}$. For the feedback of the data packet, if it is an ACK, i.e., the data packet is transmitted successfully, the nanonode can start a new data transmission or end. If it is an NAK, the nanonode will enter into the probing mode again, until the data packet is transmitted successfully or its maximum number of retransmission is reached.

A schematic diagram of the data transmission process in the ECP mechanism is shown in Figure 9.2. Note that the nanonode keeps harvesting energy in the whole process, and time synchronization is needed to maintain time consistency during network operation. Furthermore, before entering the probing mode, i.e., energy state is up on $\{E_{PT} + E_{PE}\}$, the nanonode will not transmit any probing packet and data packet. The reason is that it is useless to transmit packets if the nanonode does not have enough energy to complete the data transmission. For example, in the case of the energy stored in the nanonode is enough to transmit one probing packet and an ACK of the probing packet is received successfully. If the remaining energy of nanonode is insufficient for transmitting one data packet, and its energy-harvesting rate is very low, then it needs to take a long time to harvest enough energy γ to transmit the packet. However, the channel condition at the moment may have changed. Therefore, we consider that the nanonode transmits one data packet only after it has enough energy, i.e., its energy has to reach the energy threshold δ. Note that the current energy of nanonode can be estimated by some energy detection methods/algorithms, and the value of energy threshold in Table 9.1 can be calculated by (9.3).

9.3.2   Energy state model of the ECP mechanism

The energy distribution of nanonode can be modeled as an extended Markov process $\varphi(t)$. The process $\varphi(t)$ is represented by the Markov chain in Figure 9.1, which is fully characterized by its transition rate matrix $\mathbf{Q}$ in (9.4). In this section, we consider the behavior of the system in the steady state. Hence, matrix $\mathbf{Q}$ is provided in the steady state as well. The number of energy states in Figure 9.1 is $m = \lfloor (E_{max} - E_0)/E_{rx}^d \rfloor$, and $\mathbf{Q}$ is an $m * m$ dimensional matrix. Each element in $\mathbf{Q}$,

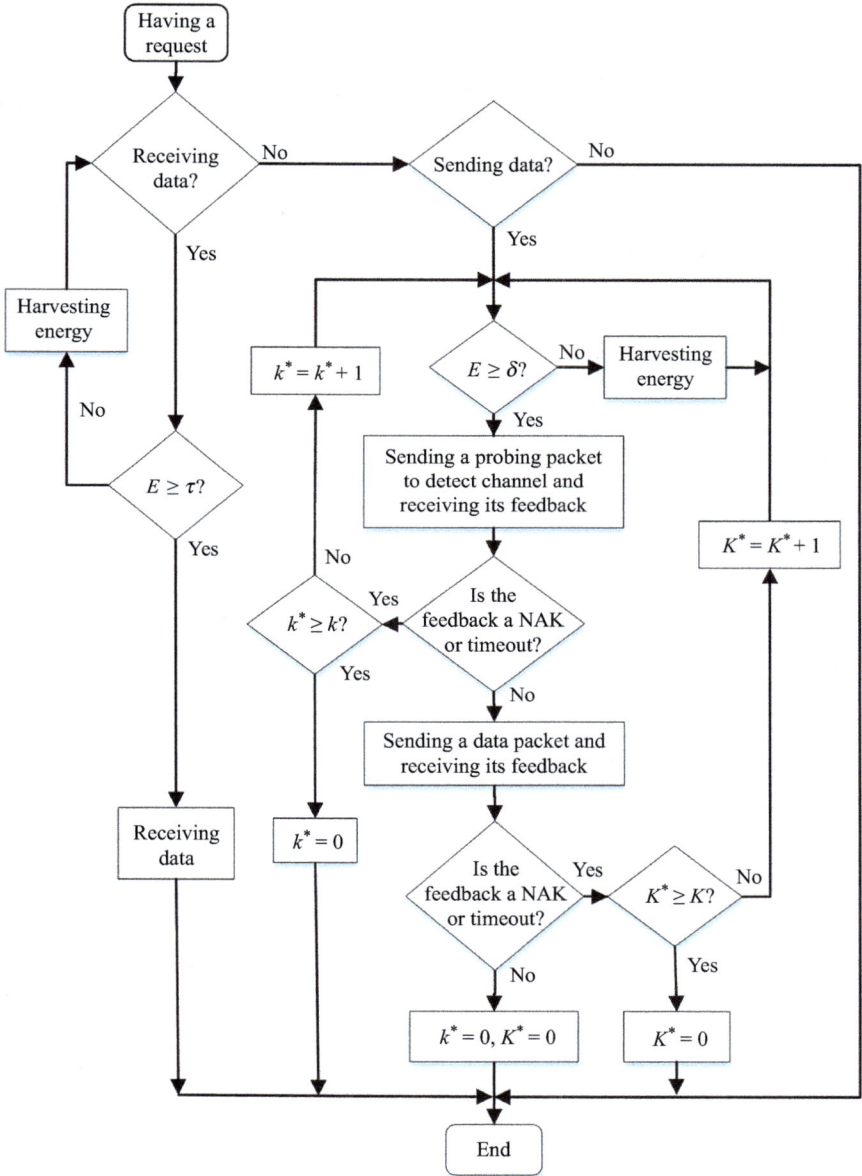

Figure 9.2   A schematic diagram of the ECP mechanism (k* and K* are the retransmission counter of one probing packet and one data packet, respectively)

e.g., q_{ij} refers to the rate of the transition from state i to state j, and $q_{ii} = -q_i = -\sum_{i \neq j} q_{ij}$. The state probability vector is defined as $\boldsymbol{\pi} = (\pi_{E_0}, \pi_{E_{PR}}, \ldots, \pi_{E_{max}})$, where π_n refers to the probability of state n in the steady state. In order to determine the parameters in matrix $\mathbf{Q}$, some definitions are necessary and given as follows:

$$
\mathbf{Q} =
\begin{array}{c}
\begin{array}{cccccccccc}
\{E_0\} & \{E_{PR}\} & \{2E_{PR}\} & \cdots & \{E_{PT}\} & \cdots & \{E_{PT}+E_{PE}\} & \cdots & \{E_{max}-E_{PE}\} & \cdots & \{E_{max}\}
\end{array} \\
\left(
\begin{array}{cccccccccc}
-\lambda_e & \lambda_e & 0 & \cdots & 0 & \cdots & 0 & \cdots & 0 & \cdots & 0 \\
\lambda_{rx} & -(\lambda_e+\lambda_{rx}) & \lambda_e & \cdots & 0 & \cdots & 0 & \cdots & 0 & \cdots & 0 \\
0 & \lambda_{rx} & -(\lambda_e+\lambda_{rx}) & \cdots & 0 & \cdots & 0 & \cdots & 0 & \cdots & 0 \\
\vdots & \vdots & \vdots & \ddots & 0 & \cdots & 0 & \cdots & 0 & \cdots & 0 \\
\lambda_{tx} & 0 & 0 & \cdots & -(\lambda_e+\lambda_{rx}+\lambda_{tx}) & \cdots & 0 & \cdots & 0 & \cdots & 0 \\
\vdots & \vdots & \vdots & \ddots & \vdots & \ddots & \vdots & \ddots & \vdots & \ddots & \vdots \\
0 & 0 & 0 & \ddots & \lambda_{px} & \cdots & -(\lambda_e+\lambda_{rx}+\lambda_{tx}+\lambda_{px}) & \cdots & 0 & \cdots & 0 \\
\vdots & \vdots & \vdots & \ddots & \vdots & \ddots & \vdots & \ddots & \vdots & \ddots & \vdots \\
0 & 0 & 0 & \ddots & 0 & \ddots & 0 & \ddots & -(\lambda_e+\lambda_{rx}+\lambda_{px}) & \ddots & 0 \\
\vdots & \vdots & \vdots & \ddots & \vdots & \ddots & \vdots & \ddots & \vdots & \ddots & \vdots \\
0 & 0 & 0 & \cdots & 0 & \cdots & 0 & \cdots & 0 & \cdots & -(\lambda_{rx}+\lambda_{px})
\end{array}
\right)
\end{array}
$$

$$(9.4)$$

First, to determine the probing packet transmission rate λ_{px}, the data packet transmission rate λ_{tx}, and the data packet reception rate λ_{rx}, the following conditions need to be satisfied:

1. One data packet cannot be transmitted if the energy level of the transmitting nanonode is lower than E_{PT}, e.g., $E_0, \ldots, E_{PT} - E_{PR}$, this probability can be given by

$$P_{drop-t}^d = \sum_{i=E_0}^{E_{PT}-E_{PR}} \pi_i \tag{9.5}$$

2. One packet cannot be received if the energy state of nanonode is $\{E_0\}$, this probability can be written as

$$P_{drop-r}^y = \pi_{E_0} \tag{9.6}$$

where $y = p$ or $y = d$ stands for the calculation of the probability of channel error for one probing packet and one data packet, respectively.

3. One packet will not be received successfully if there are errors during the transmission in the channel. This probability is obtained by

$$P_{error}^y = 1 - (1 - BER)^{N_{bits}^y} \tag{9.7}$$

where N_{bits}^y is the packet length in bits and *BER* refers to the bit error rate.

4. The packet collision probability can be calculated as a function of the network traffic λ_{net}^y, the coding weight W_y, the pulse duration time T_p, and the packet length N_{bits}^y. This probability is

$$P_{coll}^y = 1 - e^{-\lambda_{net}^y W_y T_p N_{bits}^y} \tag{9.8}$$

5. One probing packet will not be transmitted before the energy stored in nano-node reaches the energy threshold δ of the probing mode. Hence, the probability can be written as

$$P_{energy} = \sum_{i=E_0}^{\delta-E_{PR}} \pi_i \tag{9.9}$$

Based on the above analysis, the successful transmission probability of one probing packet can be given as follows:

$$P_{succ}^p = \left(1 - P_{energy}\right)\left(1 - P_{drop-r}^p\right)\left(1 - P_{error}^p\right)\left(1 - P_{coll}^p\right) \tag{9.10}$$

Similarly, the successful transmission probability of both one probing packet and one data packet can be obtained by

$$P_{succ} = P_{succ}^p \left(1 - P_{drop-r}^d\right)\left(1 - P_{error}^d\right)\left(1 - P_{coll}^d\right) \tag{9.11}$$

The network traffic rate of the probing packet λ_{net}^p between two neighboring nanonodes is defined by

$$
\begin{aligned}
\lambda_{net}^p &= \sum_{i=0}^{k} (\lambda_p + \lambda_n)\left(1 - P_{energy}\right)\left(1 - P_{succ}^p\right)^i \\
&= (M+1)\lambda_p\left(1 - P_{energy}\right)\frac{1 - \left(1 - P_{succ}^p\right)^{k+1}}{P_{succ}^p}
\end{aligned} \tag{9.12}
$$

where λ_p is the probing packet generation rate, k is the probing packet maximum number of retransmission, λ_n is the traffic from neighbor nanonodes, and we assume that $\lambda_n = M\lambda_p$, where M is the number of neighbor nanonodes.

Similarly, the network traffic rate λ_{net}^d of data packets between two neighboring nanonodes is given by

$$\lambda_{net}^d = (M+1)\lambda_d P_{succ}^p \frac{1 - \left(1 - P_{succ}^d\right)^{K+1}}{P_{succ}^d} \tag{9.13}$$

where λ_d is the data packet generation rate and K is the data packet maximum number of retransmission.

After one probing packet is transmitted successfully, the successful transmission probability of one data packet can be obtained by

$$P_{succ}^d = \frac{P_{succ}}{P_{succ}^p} \tag{9.14}$$

Then, the transmission rate λ_{tx} and the reception rate λ_{rx} of one data packet as well as the transmission rate λ_{px} of one probing packet can be written as

$$
\begin{cases}
\lambda_{tx} = \lambda_d \dfrac{1 - \left(1 - P^d_{succ}\right)^{K+1}}{P^d_{succ}} \\[2ex]
\lambda_{rx} = \lambda^d_{net} P^p_{succ} \left(1 - P^d_{drop-r}\right) \\[2ex]
\lambda_{px} = \lambda_p \dfrac{1 - \left(1 - P^p_{succ}\right)^{k+1}}{P^p_{succ}}
\end{cases}
\tag{9.15}
$$

where λ_p and λ_d are the probing and data packet generation rate, respectively. Note that the receiver counts only the packets that are not dropped in reception, and the transmitter only transmit the packets that it generates as well as it has received without errors and collision.

Finally, as shown in Figure 9.1, the transition from energy state i to state $i + 1$ happens based on the data packet energy-harvesting rate λ^i_e in energy-packet/second; the value of λ^i_e can be obtained as a function of the energy in the current state E^i_{curr} and the energy required to receive one data packet E^d_{rx}:

$$
\lambda^i_e = \frac{\lambda_{harv}\left(E^i_{curr}, E^d_{rx}\right)}{E^d_{rx}}
\tag{9.16}
$$

Note that the value of λ^i_e is changed because of the nonlinearities in the energy-harvesting process of nanonodes. The energy of the current state is obtained by

$$
E^i_{curr} = E_0 + (i - 1)E_{PR}
\tag{9.17}
$$

Up to this point, all the terms in matrix $\mathbf{Q}$ can be obtained. The steady-state probability can be determined by the transition rate matrix $\mathbf{Q}$ and the state probability vector $\boldsymbol{\pi}$ as follows:

$$
\begin{cases}
\boldsymbol{\pi}\mathbf{Q} = \mathbf{0} \\
\sum_{i \in I}\pi_i = 1, \ \pi_i \geq 0
\end{cases}
\tag{9.18}
$$

where $I = (E_0, E_{PR}, \ldots, E_{max})$.

Moreover, the probability mass function (p.m.f.) of the nanonode energy in the steady state can be computed as a function of the state probability vector $\boldsymbol{\pi}$ [24]:

$$
P_f(i) = \pi_i
\tag{9.19}
$$

where $i \in (E_0, E_{PR}, \ldots, E_{max})$, i.e., the probability that nanonodes have an energy exactly equal to i is π_i.

By the utilization of the above mathematical framework, the effect of different system parameters on the network performance can be numerically investigated.

9.4 Simulation and performance analysis

In this section, first, in order to validate the energy state model and evaluate the performance of the ECP mechanism for nanonetworks, MATLAB® is used to conduct the simulation for nanonetworks composed of 100 randomly deployed nanonodes, which transmit packets in a multi-hop fashion. The density of active nanonodes is equal to 1 node/cm². Each nanonode harvests vibrational energy by means of a piezoelectric nanogenerator with the parameters in Table 9.2. The nanobattery is fully discharged at the beginning of the simulation. Nanonodes broadcast packets to the neighboring nanonodes by means of TS-OOK. Then, the energy state model is used to investigate the impact of energy on four different common metrics in nanonetworks, and we compare the performance of the ECP mechanism with other four different error-control techniques (ARQ, FEC, EPC, and HEPC). More specifically, a 16-bit CRC is used for error detection in ECP and ARQ with a 30-bit long acknowledgment packet. For the FEC and EPC, a Hamming (15,11) code and a 12-bit low-weight code with codeword size of 15 bits are assumed, respectively, and for HEPC, the code distance D_c is set to 3. The probability of transmitting a pulse for ARQ, ECP, and FEC is 0.5, for EPC is 0.31, and for HEPC is 0.06. Note that the average energy-harvesting rate of nanonode is calculated and used in the simulations. The calculation of some other parameters about the five error-control methods can be found in [12,17,24]. The parameters used in the simulations are listed in Table 9.2.

9.4.1 Validate the energy state model of ECP

To validate the performance of the proposed ECP mechanism, we first consider the impact of packet generation rate, which follows a Poisson distribution and can be obtained by $\lambda_y = \lambda_{info}/N_{bits}^y$, where λ_{info} accounts for information generation rate and N_{bits}^y is the packet length. The separation between symbols (pulses or silences) is of 100 ps. Note that the normalized histogram of the nanonode energy state evolution over time in the simulation is calculated by the Markov Chain Monte Carlo method and is compared with the p.m.f. P_f in (9.19), obtained from the

Table 9.2 Parameters in the simulations

Parameters	Value	Parameters	Value
C_{cap}	9 nF	E_{pul-r}	0.1 pJ
V_g	0.42 V	BER	10^{-4}
t_{cyc}	1/50 s	W_d, W_p	0.5
ΔQ	6 pC	k, K	1
N_{bits}^d	194 bits	M	99
N_{bits}^p	39 bits	$E_{hold}, E_{shift}, E_{load}$	0.1 aJ
N_{bits}^f	30 bits	T_{cyc}	1 ps
T_p	100 fs	D_c	3
E_{pul-t}	1 pJ	λ_{info}	7 bits/s

proposed energy state model. Then, the probability distribution histogram of the energy state model by simulation and numeral calculations is presented. In the simulation, it takes almost 59 h to compute and draw the histogram in Figures 9.3 and 9.4. We discarded the initial samples of each run and only consider the steady state of the network. The results are shown for different information generation rates λ_{info} and bit error rates *BER*. As can be observed, the simulation results and the numerical results are matched accurately in the steady state. Moreover, it is clear that for high information generation rates or bit error rates, e.g., 9 bits/s or 10^{-2}, the center of the probability distribution is around the lower energy states, i.e., the nanonode lacks enough energy in most cases. As the information generation rate or bit error rate is decreased, the center of the probability distribution shifts toward the higher energy states.

9.4.2   Successful packet delivery probability

The end-to-end successful packet delivery probability is investigated as the first performance metric in nanonetworks, which can be defined as

$$P_{succ-e2e}^{ECP} = \left(1 - (1 - P_{succ})^{K+1}\right)^{N_{hop}} \tag{9.20}$$

where N_{hop} is the total number of hops. The end-to-end successful data packet delivery probability, $P_{d-succ-e2e}^{ECP}$, can be obtained by replacing P_{succ} in (9.20) with P_{succ}^d in (9.14). The end-to-end successful packet delivery probability, $P_{succ-e2e}$, for the ECP mechanism and the other four different error-control strategies as well as the end-to-end successful data packet delivery probability of ECP, $P_{d-succ-e2e}^{ECP}$, are shown in Figure 9.5 as a function of the data packet size N_{bits}^d. From the results, first, it is shown that the value of $P_{succ-e2e}^{ECP}$ and $P_{d-succ-e2e}^{ECP}$ are significantly higher than the other three error-control techniques (ARQ, FEC, and EPC). The reason is that in the ECP mechanism, one data packet will be transmitted only after transmitting one probing packet successfully, which improves the successful transmission probability of the data packet by detecting the condition of the channel. Moreover addition, the size of the probing packet is smaller than the data packet, so it has a higher probability to be transmitted successfully. Second, as the data packet size increases, the end-to-end successful packet delivery probability of all error-control strategies is decreased, mainly because the transmission of longer packet increases the probability of channel transmission errors and collision with the transmissions from other nanonodes given by (9.7) and (9.8). Finally, from Figure 9.5, we can find that the successful data packet delivery probability for the ECP mechanism, $P_{d-succ-e2e}^{ECP}$, is significantly improved compared with the other four schemes.

9.4.3   Delay

The end-to-end packet delay is investigated as the second performance metric in nanonetworks, which is computed as

$$
\begin{aligned}
T_{e2e}^{ECP} = N_{hop} &\left(\sum_{i=0}^{k} (2T_{prop} + T_{data}^p + T_{code}^p + T_{decode}^p + iT_{t/o}^p) \right. \\
&\left. + P_{succ}^p \sum_{j=0}^{K} (2T_{prop} + T_{data}^d + T_{code}^d + T_{decode}^d + jT_{t/o}^d) \right)
\end{aligned} \tag{9.21}
$$

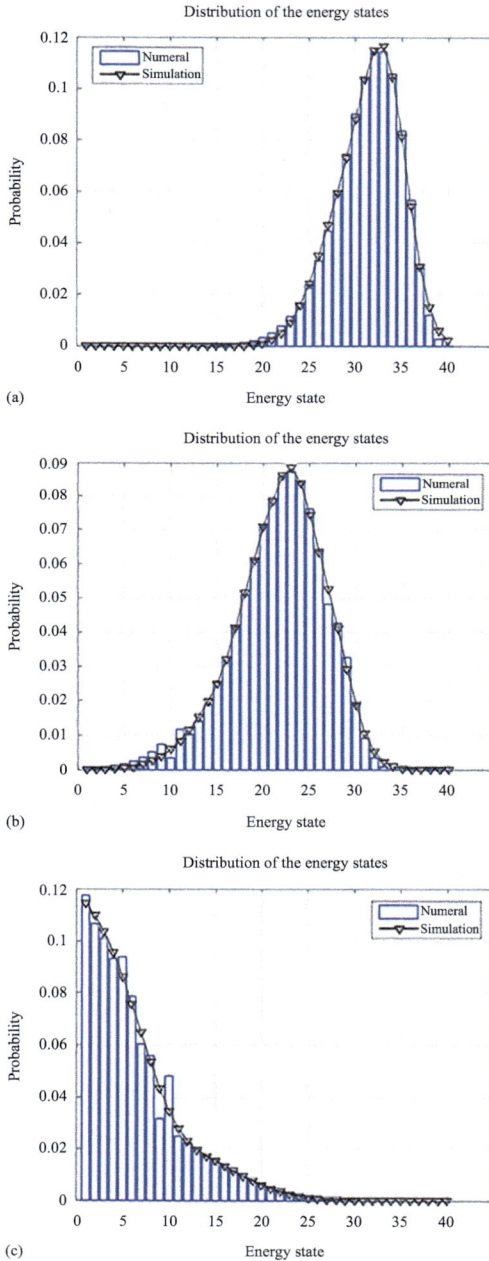

Figure 9.3   *Probability mass function p(i) of the nanonode energy in (9.19) as a function of the energy state {i} for different information generation rates λ_{info}: (a) $\lambda_{info} = 9\,bits/s$; (b) $\lambda_{info} = 7\,bits/s$; and (c) $\lambda_{info} = 5\,bits/s$*

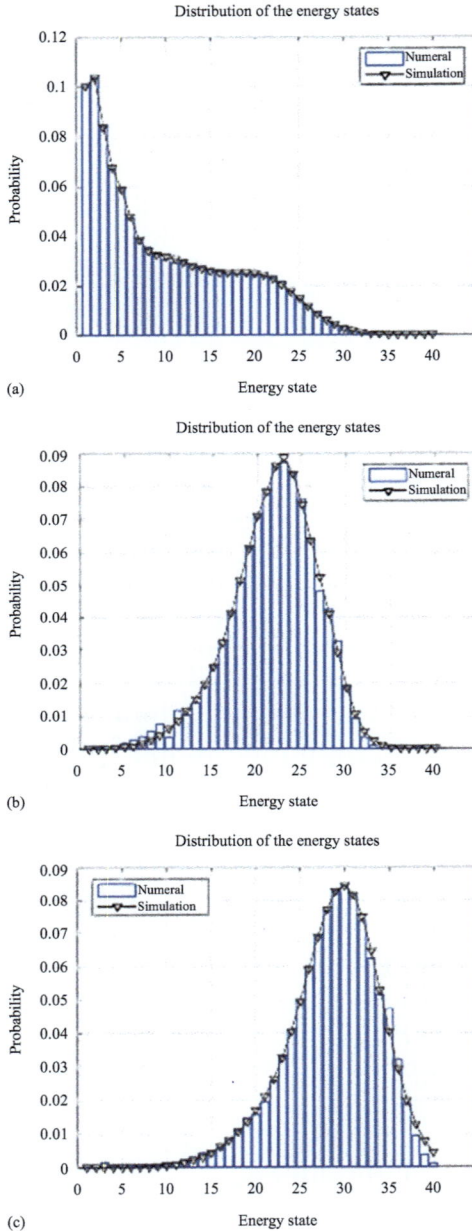

Figure 9.4   *Probability mass function p{i} of the nanonode energy in (9.19) as a function of the energy state {i} for different bit error rates BER: (a) BER = 10^{-2}; (b) BER = 10^{-4}; and (c) BER = 10^{-6}*

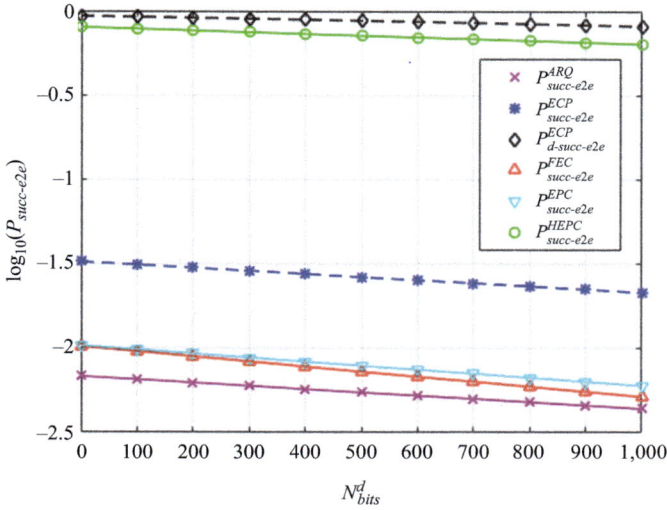

Figure 9.5　End-to-end successful packet delivery probability for the ECP mechanism (dashed lines) and other four error-control strategies (solid lines)

where T_{prop} is the propagation time. For $y = p, f$, or d, T_{data}^y stands for the different packet transmission time of one probing packet, one feedback packet, and one data packet, respectively, which can be directly obtained from the physical-layer data rate. $T_{t/o}^p$ and $T_{t/o}^d$ stand for the time out of one probing packet and one data packet, respectively. The calculations of $T_{t/o}^d$ are similar to $T_{t/o}^p$, and $T_{t/o}^p$ is defined as follows:

$$T_{t/o}^p = P_{energy} T_t^e + (1 - P_{energy})(P_{drop-r}^p T_r^p + (1 - P_{drop-r}^p)(1 - (1 - P_{error}^p)(1 - P_{coll}^p))T_o) \tag{9.22}$$

where T_t^e and T_r^p are the average time needed to harvest enough energy to reach energy threshold δ of probing mode and receive one probing packet, respectively, and are given by

$$\begin{cases} T_t^e = \sum_{i=E_0}^{\delta - E_{PR}} \pi_i / q_i \\ T_r^p = \pi_{E_0} / q_{E_0} \end{cases} \tag{9.23}$$

where q_{E_0} is the opposite number of the first element in the transition rate matrix $\mathbf{Q}$, i.e., λ_e. T_o is a random back-off time before retransmitting when the packets cannot be received correctly due to the channel errors or collisions. The coding time, T_{code}^y, introduced by the computation of a 16-bit CRC is given by [12]:

$$T_{code}^y = N_{bits}^y T_{cyc} \tag{9.24}$$

where T_{cyc} is the inverse of the clock at the nanomachine. Assume that the decoding time T_{decode}^y is equal to the coding time T_{code}^y.

The end-to-end packet delay, given by (9.21), as a function of data packet size is shown in Figure 9.6 for five different error-control methods. As can be seen, T_{e2e}^{ECP} is higher than the other four error-control schemes, i.e., the ECP mechanism needs more time than the other four error-control techniques when one data packet is transmitted from the transmitter to the receiver successfully. This is mainly because ECP needs to transmit a probing packet first, which adds extra delay but improves successful data packet delivery probability. Moreover, as the data packet size increases, the end-to-end packet delay for FEC, EPC, and HEPC increased significantly, but for ECP and ARQ, it increased slightly. The reason is that in ECP and ARQ strategies if retransmission is needed due to the lack of energy at the transmitter or the receiver, the delay is determined by the necessary waiting time $T_{t/o}^p$, which is the time to recharge the energy system up to the threshold δ for ECP and E_{PT} for ARQ. On the other hand, FEC, EPC, and HEPC can save much time

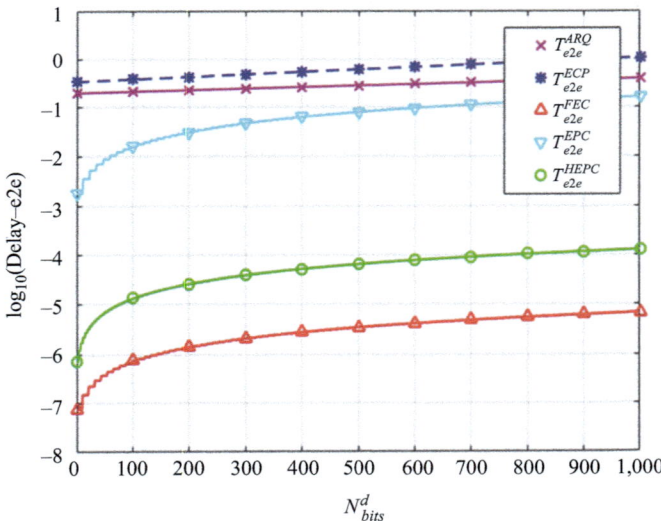

Figure 9.6   *End-to-end packet delay for the ECP mechanism (dashed lines) and other four error-control strategies (solid lines)*

because they do not need to spend a lot of time waiting to harvest enough energy to perform retransmission, especially when the energy-harvesting rate is low.

9.4.4   Throughput

The throughput is investigated as the third performance metric in nanonetworks, which can be defined as

$$th_{put}^{ECP} = \frac{P_{succ-e2e}^{ECP} \cdot N_{data}^{d}}{T_{e2e}^{ECP}} \tag{9.25}$$

where N_{data}^{d} is the number of data bits per data packet. The performance of throughput in nanonetworks with the proposed ECP mechanism, given by (9.25), and the other four error-control methods are evaluated in Figure 9.7, which is illustrated as a function of the data packet size. As shown in the figure, the throughput th_{put}^{ECP} of the ECP mechanism is lower than FEC, th_{put}^{FEC}, and HEPC, th_{put}^{HEPC}, but higher than ARQ, th_{put}^{ARQ}, and EPC, th_{put}^{EPC}. The reason is that the probing mode has a significant impact on the end-to-end successful packet delivery probability and the end-to-end packet delay, which improves $P_{succ-e2e}^{ECP}$ and $P_{d-succ-e2e}^{ECP}$ but increases T_{e2e}^{ECP}. Moreover, the throughput for ARQ and ECP is enhanced with the increase of data packet size; however, for FEC, EPC, and HEPC, the trend is opposite.

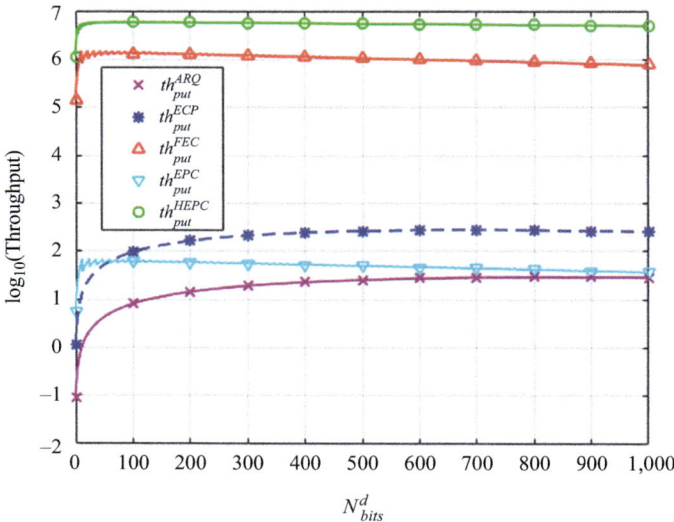

Figure 9.7   Throughput for the ECP mechanism (dashed lines) and other four error-control strategies (solid lines)

9.4.5 Energy consumption

The energy consumption of each hop is investigated as the fourth performance metric in nanonetworks, which can be defined as

$$
\begin{aligned}
E^{ECP} = P_{succ}^{-1}\big(&E_{tx}^{p} + E_{rx}^{p} + E_{tx}^{f} + E_{rx}^{f} + E_{code}^{p} + E_{decode}^{p} \\
&+ E_{code}^{f} + E_{decode}^{f} + P_{succ}^{p}(E_{tx}^{d} + E_{rx}^{d} + E_{tx}^{f} + E_{rx}^{f} \\
&+ E_{code}^{d} + E_{decode}^{d} + E_{code}^{f} + E_{decode}^{f}))
\end{aligned}
\tag{9.26}
$$

where P_{succ}^{-1} refers to the expected number of retransmissions. For $y = p, f$, or d, the coding energy, E_{code}^{y}, consumed to compute a 16-bit CRC is obtained by [12]:

$$
E_{code}^{y} = 16N_{bits}^{y}\big(E_{shift} + E_{hold}\big)
\tag{9.27}
$$

where E_{shift} and E_{hold} refer to the energy consumed to shift and hold the registry value, respectively. In our analysis, for the decoding energy consumption, we consider that $E_{decode}^{y} = E_{code}^{y}$.

In our analysis, we assume that the feedback packets can be transmitted and received successfully. Note that even if the packets (including data packets, probing packets, and feedback packets) are not correctly received because of collisions or channel errors, the energy is consumed. For the computation of energy consumption, we have taken into account both the communication energy consumption and the computation energy consumption. In Figure 9.8, the energy consumption of

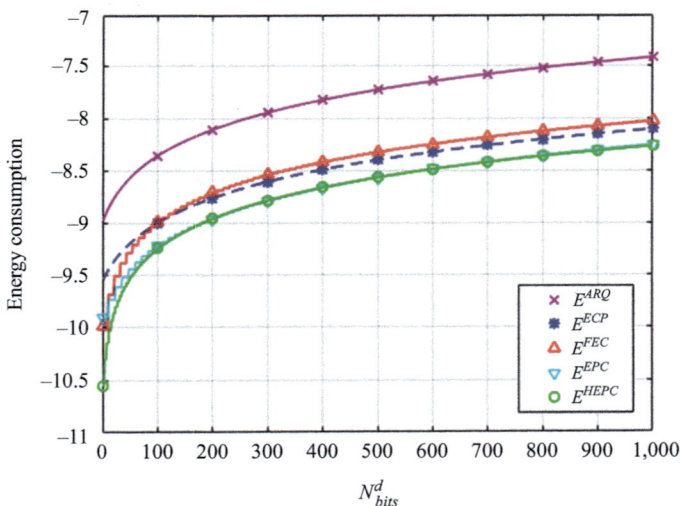

Figure 9.8   Energy consumption of successfully transmitting one data packet for the ECP mechanism (dashed lines) and other four error-control strategies (solid lines)

E^{ECP} in (26) and other error-control techniques are shown as a function of the data packet size N_{bits}^d. As shown in the figure, the trend for all error-control strategies is very similar to that of the delay shown in Figure 9.8, that is, the larger the transmitted data packet, the more the energy consumed. The energy consumption of transmitting one packet is calculated from (9.3). Alternatively, compared with the other four error-control schemes, although ECP consumes more energy than EPC and HEPC, it saves more energy than ARQ and FEC. The reason is that before one data packet is transmitted, the channel condition is known by transmitting one probing packet. Specifically, if the channel condition is poor, ECP can save more energy because it does not transmit/retransmit data packets blindly. Obviously, energy utilization can be significantly improved in nanonetworks.

9.5 Conclusion

In this chapter, first, we proposed a novel error-control mechanism for nanonetworks, and the detailed description of the mechanism is shown in Section 9.3. Second, an energy state model based on the extended Markov chain approach of the proposed ECP mechanism was presented, and the corresponding probability distribution was comprehensively investigated. Finally, we validated the energy state model and investigated the impact of the ECP mechanism on the performance of nanonetworks under different communication parameters by simulation and numeral calculations. From the above results, compared with traditional ARQ and FEC strategies, the ECP mechanism is shown to significantly improve the end-to-end successful data packet delivery probability and the energy utilization performance. Similarly, the ECP mechanism also has a better performance such as packet delivery probability and throughput than EPC strategies.

References

[1]   I. F. Akyildiz and J. M. Jornet, "Electromagnetic wireless nanosensor networks," *Nano Communication Networks*, vol. 1, no. 1, pp. 3–19, 2010.

[2]   J. M. Jornet and I. F. Akyildiz, "Femtosecond-long pulse-based modulation for terahertz band communication in nanonetworks," *IEEE Transactions on Communications*, vol. 62, no. 5, pp. 1742–1754, 2014.

[3]   I. F. Akyildiz, F. Brunetti, and C. Blázquez, "Nanonetworks: A new communication paradigm," *Computer Networks*, vol. 52, no. 12, pp. 2260–2279, 2008.

[4]   A. Cabellos-Aparicio, I. Llatser, E. Alarclon, A. Hsu, and T. Palacios, "Use of terahertz photoconductive sources to characterize tunable graphene RF plasmonic antennas," *IEEE Transactions on Nanotechnology*, vol. 14, no. 2, pp. 390–396, 2015.

[5]   J. M. Jornet and I. F. Akyildiz, "Graphene-based plasmonic nano-antenna for terahertz band communication in nanonetworks," *IEEE Journal on Selected Areas in Communications*, vol. 31, no. 12, pp. 685–694, 2014.

[6]   X. W. Yao, C. C. Wang, W. L. Wang, and J. M. Jornet, "On the achievable throughput of energy-harvesting nanonetworks in the terahertz band," *IEEE Sensors Journal*, vol. 18, no. 2, pp. 902–912, 2018.

[7]   I. F. Akyildiz, C. Han, and S. Nie, "Combating the distance problem in the millimeter wave and terahertz frequency bands," *IEEE Communications Magazine*, vol. 56, no. 6, pp. 102–108, 2018.

[8]   V. Ryzhii, M. Ryzhii, V. Mitin, and T. Otsuji, "Toward the creation of terahertz graphene injection laser," *Journal of Applied Physics*, vol. 110, no. 9, p. 094503, 2011.

[9]   T. Otsuji, S. B. Tombet, A. Satou, M. Ryzhii, and V. Ryzhii, "Terahertz-wave generation using graphene: Toward new types of terahertz lasers," *Proceedings of the IEEE*, vol. 19, no. 1, pp. 1–13, 2013.

[10]   L. Vicarelli, M. S. Vitiello, D. Coquillat, *et al.*, "Graphene field-effect transistors as room-temperature terahertz detectors," *Nature Materials*, vol. 11, no. 10, pp. 865–871, 2012.

[11]   M. Patil and R. C. Biradar, "Dynamic error control scheme based on channel characteristics in wireless sensor networks," in *IEEE International Conference on Recent Trends in Electronics, Information & Communication Technology*, pp. 736–741, 2018.

[12]   N. Akkari, J. M. Jornet, P. Wang, *et al.*, "Joint physical and link layer error control analysis for nanonetworks in the terahertz band," *Wireless Networks*, vol. 22, no. 4, pp. 1221–1233, 2016.

[13]   M. C. Vuran and I. F. Akyildiz, "Error control in wireless sensor networks: A cross layer analysis," *IEEE/ACM Transactions on Networking*, vol. 17, no. 4, pp. 1186–1199, 2009.

[14]   J. F. Cheng, "On the coding gain of incremental redundancy over chase combining," in *Global Telecommunications Conference, 2003. GLOBE-COM '03*, vol. 1, pp. 107–112, 2004.

[15]   J. M. Jornet, "Low-weight error-prevention codes for electromagnetic nano-networks in the terahertz band," *Nano Communication Networks*, vol. 5, no. 1–2, pp. 35–44, 2014.

[16]   K. Chi, Y. H. Zhu, X. Jiang, and X. Tian, "Optimal coding for transmission energy minimization in wireless nanosensor networks," *Nano Communication Networks*, vol. 4, no. 3, pp. 120–130, 2013.

[17]   P. Johari and J. M. Jornet, "Packet size optimization for wireless nanosensor networks in the terahertz band," in *IEEE International Conference on Communications*, pp. 1–6, 2016.

[18]   D. Niyato, E. Hossain, M. M. Rashid, and V. K. Bhargava, "Wireless sensor networks with energy harvesting technologies: a gametheoretic approach to optimal energy management," *IEEE Wireless Communications*, vol. 14, no. 4, pp. 90–96, 2007.

[19]   S. Sudevalayam and P. Kulkarni, "Energy harvesting sensor nodes: Survey and implications," *IEEE Communications Surveys & Tutorials*, vol. 13, no. 3, pp. 443–461, 2011.

[20] Z. L. Wang, "Towards self-powered nanosystems: From nanogenerators to nanopiezotronics," *Advanced Functional Materials*, vol. 18, no. 22, pp. 3553–3567, 2008.

[21] Y. Hu, Y. Zhang, C. Xu, L. Lin, R. L. Snyder, and Z. L. Wang, "Self-powered system with wireless data transmission," *Nano Letters*, vol. 11, no. 6, pp. 2572–2577, 2011.

[22] S. Xu, B. J. Hansen, and Z. L. Wang, "Piezoelectric-nanowire-enabled power source for driving wireless microelectronics," *Nature Communications*, vol. 1, no. 7, p. 93, 2010.

[23] S. Xu, Y. Qin, C. Xu, Y. Wei, R. Yang, and Z. L. Wang, "Self-powered nanowire devices," *Nature Nanotechnology*, vol. 5, no. 5, pp. 366–373, 2010.

[24] J. M. Jornet and I. F. Akyildiz, "Joint energy harvesting and communication analysis for perpetual wireless nanosensor networks in the terahertz band," *IEEE Transactions on Nanotechnology*, vol. 11, no. 3, pp. 570–580, 2012.

[25] J. C. Pujol, J. M. Jornet, and J. S. Pareta, "PHLAME: A physical layer aware mac protocol for electromagnetic nanonetworks," in *Computer Communications Workshops*, pp. 431–436, 2011.

[26] X. W. Yao, W. L. Wang, and S. H. Yang, "Joint parameter optimization for perpetual nanonetworks and maximum network capacity," *IEEE Transactions on Molecular, Biological and Multi-Scale Communications*, vol. 1, no. 4, pp. 321–330, 2015.

Chapter 10

Conclusion and future outlook

Akram Alomainy[1], Ke Yang[2], Xin-Wei Yao[3],
Muhammad Ali Imran[4] and Qammer Hussain Abbasi[4]

10.1 Conclusion

Nanocommunication has driven significant research interests since its proposal in 2008 as it will enable many applications in numerous fields, not only healthcare and monitoring but also industrial development and environment protection. Although the study of nano-network communication is still at its early phase, it is generally believed that the research work on the hardware-oriented research studies and communication-focused investigations on the electromagnetic (EM) communication paradigm of nano-communication is essential, which could pave the road and make the future work much more convenient.

In this book, the performances of the nano-EM communication network are thoroughly investigated with the inclusion of the overview of the current development of terahertz (THz) technologies, which is generally considered as the most promising candidate of the frequency choices. At the same time, a profound growth has been seen by the wireless interaction of the human bodies with the nano-devices in the past few years. Such growth would cause the global wearable devices market increasing from 20 million in 2015 to 187.2 million annually by 2020 [1], which also emphasizes the importance of such studies in the book.

The book starts with a simple introduction on nano-communication but emphasizing the work on nano-electromagnetic communication network in the following part. In the beginning, the elements of the wireless network are investigated where the state-of-the-art techniques related to graphene are illustrated. The devices made of such novel materials such as antenna and transceiver are investigated. Based on these research studies, the description of the network performances based on the study of channel performances, modulation schemes, coding/decoding techniques, and so on validates the capability of the nano-sensor network. The

[1]School of Electronic Engineering and Computer Science, Queen Mary University of London, London, UK
[2]School of Marine Science and Technology, Northwestern Polytechnical University, Xi'an, China
[3]College of Computer Science and Technology, Zhejiang University of Technology, Hangzhou, China
[4]James Watt School of Engineering, University of Glasgow, Glasgow, UK

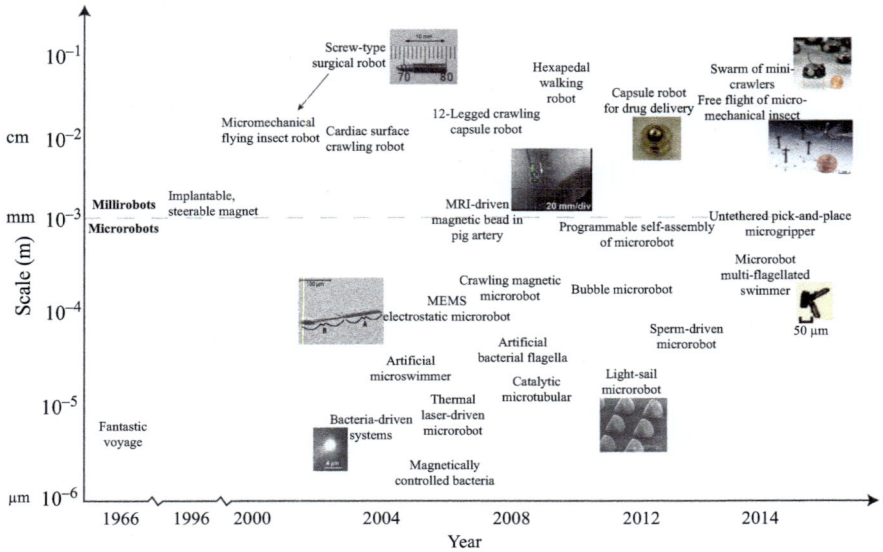

Figure 10.1   Development routine of the micro/nano-devices

study shows that more work needs to be done with the peculiarity of nano-network in mind instead of reproducing the traditional method. Meanwhile, the discussion of the antenna design with THz technology shows the promising future of the application of another novel material—Perovskite. Also, its performances for the THz body-centric network are studied. Furthermore, the research studies on the connectivity of micro- and macro-interfaces for the envisioned targeted drug delivery (TDD) system set an excellent example for the researches on the simulation platform and experimental platform, where the dependence of network structures on the applications is mentioned, pointing out the necessity of the divergent analysis methods for the thorough investigation of the nano-network.

Overall, even though the current development of the THz technology is not sufficient enough to support the whole nano-EM network, the bright future could be still seen in every chapter, especially the integration with the other communication paradigms.

10.2   Challenges and future work

With the development of the novel manufacturing techniques, the size of the sensor or machine can be made as small as micrometer, which is shown in Figure 10.1. However, the miniaturization of the nano-system is still very challenging, which needs to consider the assembly of the power feeding elements and propulsion elements. Also, the aim of the biodegradability and bio-compatibility is hard to achieve, relying on the discovery of the novel materials, but the development of which is too slow. At the same time, the current modulation schemes are not very efficient, which cannot fully take advantages of the characteristics of the THz channel. The same problems happen to the coding/decoding schemes. Furthermore, the conflict between nano-devices when communication occurs concurrently is too severe to support the successful communication link in the network. From the current study, it is found that the maximum communication distance between the single nano-devices is only at the order of meters in the open air and at the order of millimeters for the in-body communication; thus, how to make the whole communication distance long enough to be applicable is also a very challenging issue in the current study for the researches in the nano-EM fields.

In line with the challenges presented above, the future research directions that would make potential and natural progression to complete the studies in the nano-network can be summarized as follows:

10.2.1 Investigations on the novel materials

From the previous experiences, the discovery of a novel material would make the development of the engineering leaping forward by a huge gap. Taking the investigation of graphene for an example, the study of its characteristics solidates the concept of nano-EM communication with the application of THz technology, which sets up the foundation of the current work. In the book, the study on the Perovskite enables the design of the terahertz antenna, which brings the bright future of the short-range body-centric communication. It is generally believed by the author that the research studies on the biomaterials would boost the development of the truly nano-devices, enabling the realization of nano-EM networks. Also, the new material would also enable the transfer from different signal types, which would make the conversion between different communication methods possible. Furthermore, we believe that the more discoveries on the biomaterials, the closer the nano-network comes into reality.

10.2.2 Integration techniques of diverse communication methods

Currently, every communication method seems well developed in their own fields; however, the transformation between each other is still at its initial stage, especially for the nano-devices. In the book, such problems are discussed for the TDD system but only in the shallow phase. The research studies on the interfaces between the different communication methods should be deeply investigated, which are apparently not sufficient right now, not only from the material perspective but also at the device level. The same problems occur for the interfaces of macro- and micro-scenarios. It is believed that if such problems are tackled, the major problems in nano-networks would be solved.

10.2.3 Development of the common platform

Although there are a lot of communication paradigms for nano-communication, the study on the interaction between two different communications, for example, the

EM communication and the molecular communication, is still missing. It is generally believed that by emerging all the communications together, the nano-network would be much more flexible and powerful. With the comprehensive simulator proposed in IEEE P1906.1 based on the ns-3 platform in mind, different communication schemes would be studied by comparing with the EM one. At the same time, the emerging platform to simulate the hybrid communication for nano-communication should be studied. Also, from the study of the models of nerve system and skin, it seems dispensable to study the detailed models when the size of the functional devices goes down to the milli/nano-scale; therefore, the platform to simulate the interaction between the environment and the devices should be further studied to make sure the whole system functions normal.

10.2.4   Introduction of big data analysis techniques

The integration of big data techniques with the nano-network would be a hot topic because of the nature of nano-network that numerous stakeholders are involved in data generation and management. From the perspective of the data analysts, the study is still missing. First, the standardization of the data format and protocols should be set up while a unified data schema should be put forward and adopted by the whole network investigators. Most importantly, new powerful analytic tools should be developed because the amount of data we will face would go up exceedingly.

10.2.5   Security issues

The security issues contain two parts: the study of the effects of the nano-devices on the human body and the study of the data leakage of the nano-network. For the first issue, the research studies on the biomaterials should be carried out with the emphasis on the investigation of the influences on the human body. Since the nano-devices do not have much processing power, data security becomes a challenging task against data hacking and malicious adversary attempts; thus, new robust techniques with a simple realization method are required to ensure data security and privacy, which makes such work tough.

Reference

[1]   Wearable device market forecasts. Boulder, CO: Tractica; 2015.

Index